Challenges of the Unseen World

A LABORATORY COURSE IN MICROBIOLOGY

Challenges of the Unseen World

A LABORATORY COURSE IN MICROBIOLOGY

Richard J. Meyer, PhD
The University of Texas at Austin
College of Natural Sciences
Austin, Texas

Stacie A. Brown, PhD
Director of First Year Biology Laboratories
Southwestern University
Georgetown, Texas

WASHINGTON, DC

Disclaimer: To the best of the publisher's knowledge, this publication provides information concerning the subject matter covered that is accurate as of the date of publication. The publisher is not providing legal, medical, or other professional services. Any reference herein to any specific commercial products, procedures, or services by trade name, trademark, manufacturer, or otherwise does not constitute or imply endorsement, recommendation, or favored status by the American Society for Microbiology (ASM). The views and opinions of the author(s) expressed in this publication do not necessarily state or reflect those of ASM, and they shall not be used to advertise or endorse any product.

Library of Congress Cataloging-in-Publication Data

Names: Meyer, Richard J. (Professor of molecular biosciences), author. | Brown, Stacie A., author.
Title: Challenges of the unseen world : a laboratory course in microbiology / Richard J. Meyer, Stacie A. Brown.
Description: Washington, DC : ASM Press, [2018] | Includes bibliographical references.
Identifiers: LCCN 2017055532 | ISBN 9781555819927 (print : alk. paper)
Subjects: LCSH: Microbiology—Laboratory manuals. | Microbiology—Study and teaching.
Classification: LCC QR63 .M49 2018 | DDC 579.078—dc23
LC record available at https://lccn.loc.gov/2017055532

Printed in the United States of America

10 9 8 7 6 5 4 3 2 1

Address editorial correspondence to
ASM Press, 1752 N St., N.W.,
Washington, DC 20036-2904, USA

Send orders to ASM Press, P.O. Box 605, Herndon, VA 20172, USA
Phone: 800-546-2416; 703-661-1593
Fax: 703-661-1501
E-mail: books@asmusa.org
Online: http://www.asmscience.org

Cover and interior design: Susan Brown Schmidler

Cover illustration:
Confocal micrograph of the bacteria *Escherichia coli (E. coli).* Fluorescent proteins are being used to understand the underlying mechanisms of bacterial biofilms formation. Credit: Fernan Federici & Jim Haseloff, Wellcome Images CC BY 4.0

To our families

RICHARD J. MEYER
STACIE A. BROWN

Contents

Lab Two

Lab Three

challenge Two

Confirming the identification of a microorganism by sequencing the 16S rRNA gene 69

Preface

Why write this lab manual? This project has been fueled by two observations. The first is that the content of many microbiology lab courses has increasingly lagged behind the principles and methods emphasized in lectures and textbooks. Second, some of these lab courses cling to the traditional method of exposing students to techniques as a progression of disconnected exercises. As a result, students learn many ways to characterize microorganisms, but not how to use these tools in an integrated way to solve problems in microbiology. These limitations can be found in the popular, commercial lab manuals and in many "in-house" manuals as well.

It is not hard to understand why laboratory courses in microbiology evolve slowly. First, there is often a lack of resources. Even small changes can result in the need for new equipment, large and small, and the cost can easily exceed the budget of many departments. These budgets are often prepared under the assumption that they will be stable from year to year, with little thought to the fact that a step up to more current content might require a one-time step-up in budget. Second, instructors often lack the time needed to test and then integrate new material into a lab course. In fact, there may be a disincentive for doing so: instructors are frequently evaluated by the "success" of their courses, meaning high enrollments and favorable student ratings, rather than by their content.

These problems can be less severe at well-endowed institutions offering teaching relief for course development and larger budgets. Many of these institutions produce their own lab manuals with up-to-date content. However, this content often reflects the interests and resources of the department and does not migrate well to other institutions, particularly small colleges.

We have tried to create a lab manual that offers a route to developing new content and pedagogy while remaining practical for the many undergraduate institutions that include a microbiology lab in the curriculum. We have sought to offer a course with the following characteristics.

1. The content conforms to the goals outlined in " Recommended Curriculum Guidelines for Undergraduate Microbiology Education" (American Society for Microbiology, 2012).
2. Work is arranged around solving real-world problems (a "Challenge"), with an emphasis on cooperative effort by the class.
3. Techniques are introduced as they are needed, so that students can recognize their usefulness in working toward the solution for each Challenge.
4. Background sections are included to ensure that students have the information needed to understand what they are doing and how it is related to broader concepts in microbiology.
5. The course is largely modular. Instructors can extract one or more Challenges and integrate these into their own course.
6. The modules are amenable to various modes of in-depth assessment, including individual formal lab reports and group oral presentations.
7. The manual has an Introduction that provides the foundation for the proper practice of microbiology by including:
    a. A description of the scientific method
    b. An introduction to experimental design
    c. Instructions for keeping a proper notebook
    d. Safety guidelines for BSL1 (biosafety level 1) teaching laboratories outlined by the American Society for Microbiology (ASM)
8. All the organisms for the course are BSL1, in accordance with ASM guidelines. Conditions for handling BSL2 strains are not required.
9. Much of the necessary equipment can be found in microbiology teaching laboratories or borrowed from other lab courses.
10. Questions, ranked according to Bloom's taxonomy, are provided for the lab sessions.
11. An accompanying *Handbook for the Instructor* covers in detail recipes, setup, and logistics.

Along the way it was also necessary to make some compromises:

1. The Challenges are based on situations found in the fields of medical microbiology and epidemiology. These areas engage students, many of whom are preparing for health-related careers, and they make up most of the content of many lab courses. Our emphasis is on problem-solving rather than on demonstrating the different areas of microbiology. The skills learned in the course, however, are broadly applicable to these other areas.
2. The steps to solving each challenge will differ somewhat from those used by professional medical microbiologists or epidemiologists. The differences

reflect the need to include certain fundamental techniques and to organize the work so that it fits within lab periods.

3. The work is based on a particular set of strains. These strains were selected to demonstrate different microbial characteristics while keeping the collection within a reasonable size. Where substitutions are possible, this is indicated in the *Handbook for the Instructor*. Many colleges and universities have large collections of BSL1 and BSL2 strains for teaching purposes. In some cases, these collections extend back decades and their provenance is incompletely or inaccurately recorded. We think it is safer and more convenient to work with a small number of BSL1 strains.

We thank Elizabeth Emmert, Rachel Horak, Brooke Jude, Peter Justice, and Susan Merkel for their comments on an early draft of this book. Their criticisms helped point us in the right direction for producing something worthwhile. If we still managed to lose our way, the fault is ours. Thanks also to Annie Hollingshead and Tina Shay, who maintain the microbiology teaching labs at the University of Texas at Austin. Their knowledge and experience in making new experiments practical for large groups of undergraduates helped us keep our feet on the ground.

About the Authors

Richard J. Meyer, PhD, is professor emeritus in the Department of Molecular Biosciences at the University of Texas at Austin. He joined the Department of Microbiology at the University of Texas at Austin in 1978, and has been at that institution ever since. From the beginning of his career, Meyer has been interested in the hands-on aspect of teaching biology to undergraduates. He developed the introductory microbiology laboratory course currently used at the University of Texas at Austin. The pedagogical approaches in that course inspired him to develop the manual you hold in your hands. Over more than forty years, Meyer's research was on the molecular mechanisms of replication and conjugative transfer of broad host-range plasmids.

Stacie A. Brown, PhD, is director of first year biology laboratories and a member of the Biology Department at Southwestern University. Prior to her current position, she taught microbiology courses for biology majors and pre-nursing students while also overseeing the microbiology labs at Texas State University. For several years, she also taught microbiology labs and courses at the University of Texas at Austin. Her experience teaching microbiology labs to thousands of undergraduates ensures that the challenge-based microbiology labs in this manual will work in any introductory laboratory course in undergraduate microbiology.

Introduction

This course is made up of six challenges. Each challenge contains a different problem, one that you might encounter as a microbiologist and be asked to solve. To do so, you will draw upon different techniques learned during the course, obtain and analyze the data you need, and then present the solution to the class or to your instructor.

The Scientific Method

Scientists make new observations about the world and then provide an explanation for these observations. This sounds simple, but it isn't. For one thing, a new explanation must be viewed in the context of what has already been learned. Most often, the explanation is an extension or refinement of an earlier explanation. The new explanation is more powerful because it includes more observations, but is consistent with previous thinking. Occasionally, though, the new explanation is completely different from what was thought before. When that happens, it is an exciting moment in science.

How do we go from observation to explanation? The logical structure that scientists use, consciously or not, is called the **scientific method**, outlined informally in Fig. I-1. Scientists make careful observations and then identify those that need an explanation. They learn what is already known and then propose a **hypothesis**, a tentative explanation. The hypothesis must explain the new observation while being consistent with prior observations. In addition, it must be testable. This means that if the hypothesis is true, it will lead to predictions that can be tested by experiment. Scientists design and carry out these experiments and then ask whether the results match the predictions expected from the hypothesis. If they do not, the hypothesis is discarded and a new hypothesis accommodating these results is put forward.

The application of the scientific method as a series of steps is not always obvious from the course of scientific research and discovery. However, it still forms the logical

Make observations
Which observations need to be explained?
Develop an hypothesis
Do experiments to test hypothesis
Use results to evaluate hypothesis

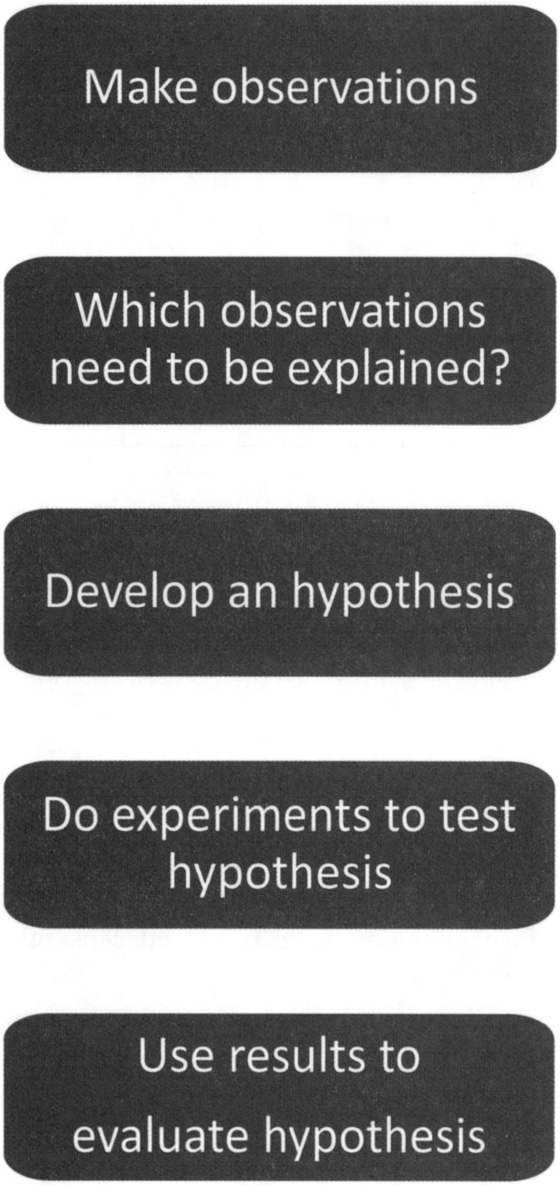

Figure I-1 The steps in the scientific method.

underpinning of how scientists approach a problem. An example is the discovery that DNA is the carrier of genetic information (Fig. I-2). In 1928, Fred Griffiths, working with the bacterium *Streptococcus pneumoniae*, discovered that if you injected a mouse with dead cells of a virulent (disease-causing) strain, along with living cells of a strain that did not cause disease, the mouse developed an infection and died. By themselves, neither the dead virulent cells nor the living avirulent cells had this effect. Griffiths concluded that a "transforming principle" from the dead cells was converting the living cells to virulence. This was exciting because the acquired virulence was stably maintained as the cells grew and divided, indicating that the virulence trait was due to inherited genes. In other words, genetic information had passed from the dead cells to the living cells.

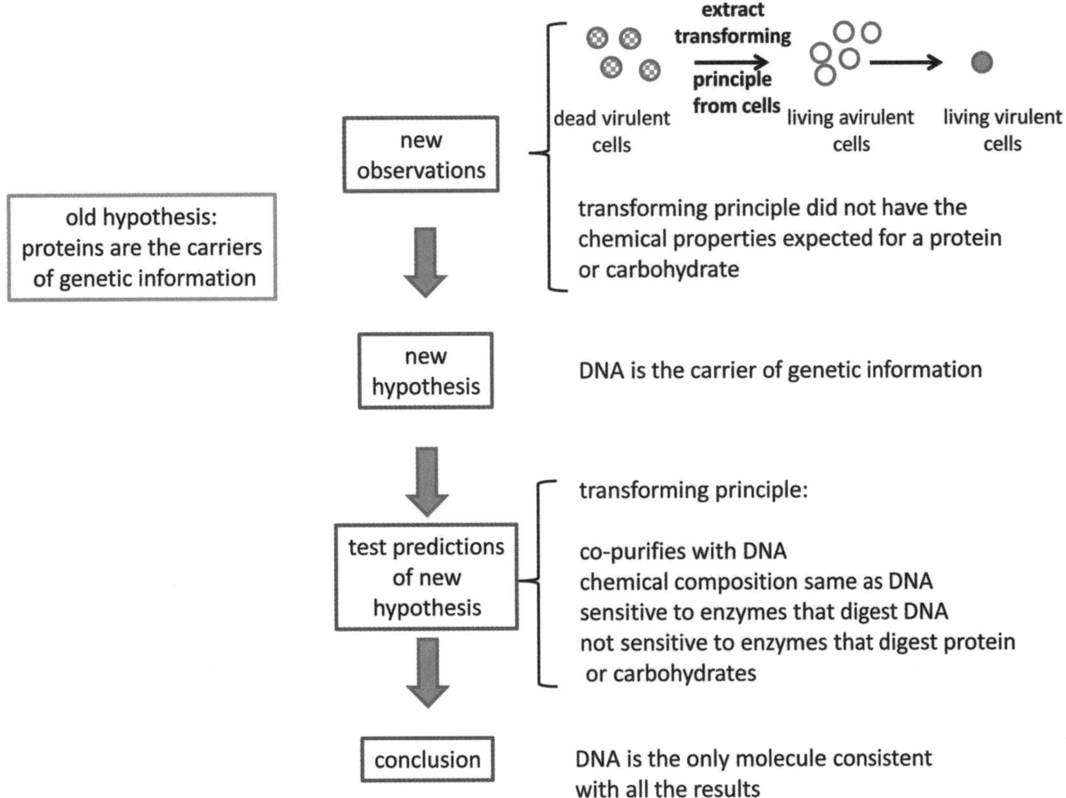

Figure I-2 Research leading to the discovery that the "transforming principle" is DNA.

What was the carrier of the genetic information? Since whole cells were used in the Griffiths experiment, there were many possibilities, but most of the bets were on proteins being the "transforming principle." The reason was that only proteins were thought to be sufficiently complex and various in their properties to convey genetic information. A critical *new observation* was provided by Dawson and Sia, who showed that the transforming substance could be extracted as a soluble component from the virulent cells and then used to transform the avirulent strain in a test tube. This meant that it might be possible to purify the transforming principle and determine its chemical properties, a fact that was recognized by Oswald Avery and his laboratory group. The first attempts at characterization indicated that it was not a protein; rather, the properties were consistent with deoxyribose nucleic acid, or DNA, another and surprising new observation. Avery and his colleagues set about testing the hypothesis that DNA was the transforming principle. If the hypothesis was true, then it would lead to several predictions that could be tested by experiment. In every case, the experimental results were consistent with Avery's hypothesis (Fig. I-2), resulting in a startling paper published by the Avery group in 1944. The idea that DNA was the carrier of genetic information was so unexpected that even Avery himself was reluctant to draw that conclusion, although, as you know, it has stood the test of time. A good review of this transformative moment in microbiology (no pun intended) is Cobb (2014).

There is an important but subtle logic behind scientific experiments. Science basically works by the process of elimination. Different hypotheses are tested by experiment and are discarded if the experimental results are inconsistent with the hypothesis. For example, the hypothesis that protein was transforming the cells was eliminated by the biochemical properties of the transforming principle. A hypothesis comes to be accepted when it is consistent with all the experimental results and when all other reasonable competing hypotheses have been ruled out by experimentation. "Reasonable competing hypotheses" depend on both our state of knowledge and our imagination. An awareness of this might have been one reason Avery was cautious about drawing the firm conclusion, in public at least, that DNA is the genetic material.

Experimental Design

From the foregoing it must be obvious that good experiments are the keystone of the scientific method. In designing an experiment, there are some things to keep in mind.

1. *Does the experiment test the hypothesis?* The purpose of a good experiment is to discriminate between hypotheses. Results that would be consistent with all the hypotheses under consideration do not help us to decide between them.

2. *Is the experiment well controlled?* Controlling all the possible variables except the condition you want to test is the best situation. In reality, this is not always possible, and there are often uncontrolled variables, variations in the experimental conditions in addition to what you want to test. Repeating the experiment multiple times, along with statistical methods, can sometimes be helpful when dealing with uncontrolled variables. However, statistical analysis cannot rescue experiments where the results are overwhelmingly influenced by the effects of uncontrolled experimental conditions.

3. *Is the sample size large enough, and can the experiment be replicated?* Sample size and replication of the experiment by yourself or others are closely related. Sometimes a result that seems to be real at first disappears upon replication. This is usually because the sample size was too small to begin with, and apparently real differences were just the result of chance. Suppose you hypothesize that a coin is weighted so that it will come up "heads" more often than "tails" after tossing. You decide to do the experiment of tossing the coin 8 times. If "heads" is the result 6 or more times, then you will conclude that your hypothesis is correct and you will publish your result. You get 7 heads and 1 tail during the toss, strong evidence, it seems, of a bad coin. However, while the probability of getting this particular result with a fair coin is only 3%, the probability of getting 6 heads or more is 14%. Your criterion for a bad coin would be met by a fair coin 14% of the time. The solution is to repeat the coin toss multiple times, which might seem obvious. However, many published experiments have not been sufficiently replicated. The inability to repro-

duce experimental results has become a major concern in the scientific community (Anonymous, 2016).

4. *Are the accuracy and precision of your measurements adequate to support your conclusion?* **Accuracy** refers to the closeness of a measurement to the true value, while **precision** refers to the reproducibility of a measurement: how often repeated measurements will give the same value. Both must be taken into consideration when drawing conclusions from an experiment. For example, a small but real change due to different experimental conditions might not be detected if the measurements are inaccurate. Imprecise measurements, on the other hand, could result in the real change becoming obscured by the random "scatter" of different data points.

5. *Could observer bias influence the results?* When scientists do experiments, they often have a desired result in mind, usually the one that supports their favorite hypothesis. This can lead to the unrecognized manipulation of results to favor this hypothesis. Sometimes rationalizations like "This value is much smaller than the rest: obviously there was a procedural error so it should be discarded" are used as a justification. This is a particular problem with students in lab classes. Often they think they know the expected outcome of an experiment. If some measurements do not support this result, they immediately assume that these were due to experimental errors and can be discarded.

For practice, consider the following situation:

A marine microbiologist suspects that iron in seawater stimulates the activity of a particular enzyme in the microbe she is studying. She takes eight samples of the seawater over the course of a month and adds the same amount of bacteria to each when she is ready to do the experiment. To four of these she also adds iron dissolved in seawater, and to the other four the same volume of seawater without iron. She then extracts the enzyme from the cells in each sample and assays the activity. The results are shown in Table I-1.

Table I-1 Effect of iron on enzyme activity

Seawater sample	Iron added	Seawater sample	No iron added
1	132	5	67
2	208	6	105
3	88	7	99
4	102	8	152
average	133		106

1. Does this experiment test the hypothesis?
2. The scientist is careful to use seawater as a control for the addition of iron. Does this mean that the experiment is well controlled? How could the experiment be better designed for stricter controls?
3. From the average result in each case, would you conclude that iron stimulates the activity of the enzyme? What is the underlying problem with these data? What changes would you make for a more convincing result?

Big Data

While the scientific method has been a fundamental rubric in research for many years, another way of making sense of the natural world has gained attention. This has come about as a consequence of "big data," databases containing vast amounts of information. In biology, this includes most famously the databases of DNA sequences for different organisms. One of these, **GenBank**, contains more than 189,232,925 sequences, made up of 203,939,111,071 bases (as of June 2016) (Sarkar, 2016). Among these data are the complete sequences of all the DNA from each of more than 13,000 different types of bacteria (https://www.ncbi.nlm.nih.gov/genbank/). The database of DNA sequences of organisms is not the only large database: for example, catalogs of all the proteins encoded by an organism and their abundance in the cell are also being created (Sarkar, 2016), as are all the metabolic pathways.

How does one begin to make sense of the huge volume of biological data being generated? Some people have argued that an approach different from the scientific method is needed. Indeed, it has been claimed that the scientific method is obsolete and that it is no longer necessary to make predictions and then test these by experiment (Anderson, 2008). Instead, huge computers should be used to sift through the data and look for correlations. Hypotheses about causation are no longer necessary, because the database is so large that a correlation is bound to be significant.

While looking for correlations is useful, it has been sharply criticized as a platform for scientific discovery (Barrowman, 2014). As pointed out in this article, larger data sets will result in a larger number of meaningless or misleading correlations rather than fewer. It will still be necessary to sort out which correlations are significant. The credo "Correlation does not imply causation" applies, regardless of the size of the data set.

The scientific method is embedded in the solutions to the challenges in this course. Before you start each challenge, frame the challenge in terms of new information, hypothesis, and testing the hypothesis.

Documentation

How your results will be presented and evaluated will be up to your instructor. However, all scientific study includes a notebook.

A lab notebook is a scrupulously honest and complete record of what you did and the results obtained. Unlike other forms of scientific presentation, doing the work and observing the results is done *at the same time* as entries into the notebook.

Many students are tempted to record their activities haphazardly on pieces of paper, or trust their memory, and then transfer the information to a lab notebook later. This is particularly true if the instructor will view their notebooks during the course and the students are eager to provide polished results. However, this is not a notebook, because entries were made at a later time than the work. This practice is more likely to result in recording errors and increases the possibility that data will be altered to fit expectations, unconsciously or not. Notebooks do not have to be neat to contain scrupulously recorded, important information (Fig. I-3).

Important practices:

1. Use a permanently bound notebook. Loose-leaf notebooks often result in lost pages.
2. Write in ink.
3. Include the date with each entry.

Figure I-3 Notebook of Albert Schatz, a graduate student in the laboratory of Selman Waksman at Rutgers in the 1940s. When carefully read, the notebook shows that it was Schatz who discovered the antibiotic streptomycin, not Waksman, who for many years received all the credit. From the Rutgers archives.

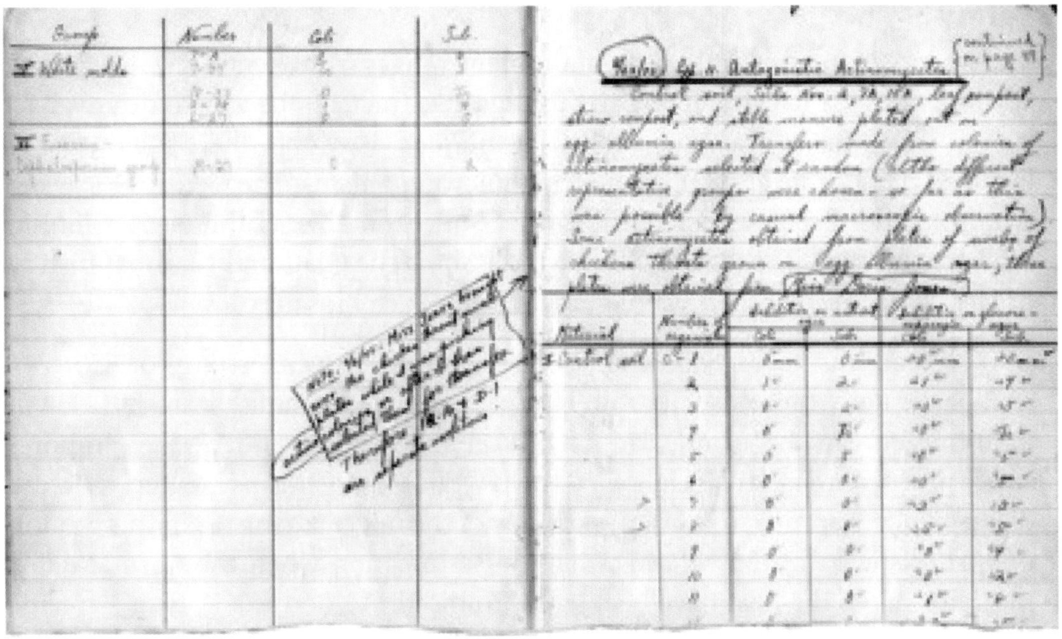

4. If you make a correction, draw a line through the corrected entry (do not erase or make illegible) and then add the correction with the date.
5. For items that need to be added, such as photographs, apply them with tape or a paper glue. Do not keep them loose in the notebook. In addition, date the item and provide a legend or labels as appropriate. When it is time to present your data, you want to be completely clear what the inserted item is showing.
6. Include important calculations in the notebook. What is an important calculation? If you are dividing 500 ml of buffer into two equal amounts, you don't have to show the calculation 500/2 = 250, but calculations critical for the success of the experiment should be recorded, for example, calculating the weight of a chemical that would be required to make 100 ml of a 0.01 M solution, or the volume of a 0.05 M stock solution needed to make 100 ml of a 0.01 M solution.

If you work in any area of science, your notebook might become very important. A dramatic example is the ongoing battle over who owns the rights to the CRISPR (clustered regularly interspaced short palindromic repeats) system for genetic engineering (Regalado, 2015). This technique, which allows the precise modification of DNA in eukaryotic cells, was developed by two different lab groups (at least). The institutions that house these groups are both claiming patent rights for a process that could earn millions or even billions of dollars in licensing fees. The patent office has decided that whoever developed the procedure first will be awarded the patent. Thus, it will all come down to the notebooks of the different scientists: who did what first?

Develop the habit of keeping a good notebook. Your instructor may ask to see your notebook or assign notebook checks for part of your grade.

Safety

The impact of different microbes on humans ranges from beneficial, as is the case with many of the bacteria found in our digestive tract, all the way to deadly. Bacteria known to cause disease are called **pathogens**. However, there is no clear line between pathogenic and nonpathogenic microorganisms. The Centers for Disease Control and Prevention (CDC) categorizes microorganisms and viruses into four biosafety levels: BSL-1, BSL-2, BSL-3, and BSL-4. The CDC and National Institutes of Health have established safety guidelines for each biosafety level.

BSL-1 organisms pose minimal risk to users. Nonpathogenic strains of *Escherichia coli*, particularly those used as model organisms in laboratories, are BSL-1. Work with these organisms can be carried out on an open bench. However, important safety practices must still be followed (below). BSL-2 organisms are known to cause disease under certain circumstances. Many of the microorganisms in this group are opportunistic pathogens; that is, they cause infection when the normal defense mechanisms of the host have been compromised. A good example of a microorganism in this group is *Pseudomonas aeruginosa*. Someone with a weakened immune system or a damaged physical barrier such as burned skin is vulnerable to infection by an opportunistic pathogen. These organisms require additional precautions, including stricter lab access and special containment equipment. Organisms in the BSL-3 category are serious pathogens that are easily spread. West Nile virus and *Mycobacterium tuberculosis*, the causative agent of tuberculosis, are examples of BSL-3 pathogens. Laboratories working with BSL-3 organisms have specialized ventilation systems, and personnel must wear additional layers of protective clothing, including gowns. Some organisms, like the Ebola virus, cause disease for which there is no vaccine or treatment. These pathogens are assigned to BSL-4. There are only a handful of BSL-4 laboratories across the country. Access to these labs is restricted, and all entry and exit is monitored. Highly trained personnel are required to work in a containment suit, and everything, including the air, must be decontaminated before it leaves the building.

Your instructor has a complete list of the microorganisms used in this lab. These organisms are all classified as BSL-1 and pose little risk to healthy individuals. If you are pregnant or immunocompromised or live with someone who is immunocompromised, take the list of organisms to your health care professional, who can advise you on the appropriate level of participation in the lab.

Despite the low risk associated with BSL-1 organisms, it is still important to follow the safety guidelines for teaching laboratories outlined by the American Society for Microbiology (ASM). The ASM guidelines for handling BSL-1 organisms are below, **with the shaded text especially important for students**. After reading the entire document, you will be asked to sign the safety contract provided by the instructor. This contract constitutes an agreement that you will conscientiously follow BSL-1 guidelines in the laboratory.

BIOSAFETY LEVEL 1 (BSL1) GUIDELINES FOR TEACHING LABORATORIES.

Preamble: Educators need to be aware of the risks inherent in using microorganisms in the laboratory and must use best practices to minimize the risk to students and the community. The following guidelines are designed to encourage awareness of the risks, promote uniformity in best teaching practices, and protect the health and wellness of our students. These guidelines are not mandatory, but are designed to promote best practices in the teaching laboratory. Note that not all institutions are equipped to handle organisms in a BSL2 setting. Work with microbes at the K–12 level, informal education settings (e.g., science fairs, museums, science centers, camps, etc.), and in undergraduate

Biosafety Level 1 continues on next page

non-microbiology laboratories would almost always be at BSL1. Even though organisms manipulated in a BSL1 laboratory pose a low level of risk to the community and are unlikely to cause disease in healthy adults, most of the microorganisms used in the microbiology teaching laboratory are capable of causing an infection given the appropriate circumstances. Many best practices should be adopted to minimize the risk of laboratory-acquired infections and to train students in the proper handling of microorganisms. The practices set forth in these guidelines fall into six major categories: personal protection, laboratory physical space, stock cultures, standard laboratory practices, training, and documents. For ease of use, the requirements and practices are brief. Explanatory notes, sample documents, and additional resources can be found in the appendix.

Personal Protection Requirements

- Wear safety goggles or safety glasses when handling liquid cultures, when performing procedures that may create a splash hazard, or when spread plating.
- Wear closed-toe shoes that cover the top of the foot.
- Wear gloves when the student's hands have fresh cuts or abrasions, when staining microbes, and when handling hazardous chemicals. Gloves are not required for standard laboratory procedures if proper hand hygiene is performed. Proper hand hygiene involves thorough hand cleansing prior to and immediately after finishing handling microorganisms and any time that microbes accidentally contact the skin. Hand cleansing is performed by washing with soap and water or rubbing with an alcohol-based hand sanitizer.
- *Recommended: Wear laboratory coats.*

Laboratory Physical Space Requirements

- Require all laboratory space to include:
  - o Nonporous floor, bench tops, chairs, and stools.
  - o Sink for hand washing.
  - o Eyewash station.
  - o Lockable door to the room.
- Follow proper pest control practices.
- *Recommended: Keep personal belongings in an area separate from the work area.*
- *Recommended: Use a working and validated autoclave.*

Stock Culture Requirements

- Only use cultures from authorized, commercial, or reputable sources (e.g., an academic laboratory or state health department). Do not subculture unknown microbes isolated from the environment because they may be organisms that require BSL2 practices and facilities.
- Maintain documents about stock organisms, sources, and handling of stock cultures.
- Obtain fresh stock cultures of microorganisms annually (e.g., purchased, revived from frozen stock cultures, etc.) to be certain of the source culture, minimize spontaneous mutations, and reduce contamination.

Standard Laboratory Practices

- Wash hands after entering and before exiting the laboratory.
- Tie back long hair.
- Do not wear dangling jewelry.
- Disinfect bench before and after the laboratory session with a disinfectant known to kill the organisms handled.
- Use disinfectants according to manufacturer instructions.
- Do not bring food, gum, drinks (including water), or water bottles into the laboratory.
- Do not touch the face, apply cosmetics, adjust contact lenses, or bite nails.
- Do not handle personal items (cosmetics, cell phones, calculators, pens, pencils, etc.) while in the laboratory.
- Do not mouth pipette.
- Label all containers clearly.
- Keep door closed while the laboratory is in session. Laboratory director or instructor approves all personnel entering the laboratory.
- Minimize the use of sharps. Use needles and scalpels according to appropriate guidelines and precautions.
- Use proper transport vessels (test tube racks) for moving cultures in the laboratory, and store vessels containing cultures in a leak-proof container when work with them is complete.
- Use leak-proof containers for storage and transport of infectious materials.
- Arrange for proper (safe) decontamination and disposal of contaminated material (e.g., in a properly maintained and validated autoclave)

Biosafety Level 1 continues on next page

or arrange for licensed waste removal in accordance with local, state, and federal guidelines.

- Do not handle broken glass with fingers; use a dustpan and broom.
- Notify instructor of all spills or injuries.
- Document all injuries according to school, university, or college policy.
- Use only institution-provided marking pens and writing instruments.
- Teach, practice, and enforce the proper wearing and use of gloves.
- Advise immune-compromised students (including those who are pregnant or may become pregnant) and students living with or caring for an immune-compromised individual to consult physicians to determine the appropriate level of participation in the laboratory.
- *Recommended: Keep note-taking and discussion practices separate from work with hazardous or infectious material.*
- *Recommended: Use microincinerators or disposable loops rather than Bunsen burners.*

Training Practices

- Be aware that student assistants may be employees of the institution and subject to OSHA, state, and/or institutional regulations.
- Conduct extensive initial training for instructors and student assistants to cover the safety hazards of each laboratory. The institution's biosafety officer or microbiologist in charge of the laboratories should conduct the training.
- Conduct training for instructors whenever a new procedural change is required.
- Conduct training for student assistants annually.
- Require students and instructors to handle microorganisms safely and responsibly.
- Inform students of safety precautions relevant to each exercise before beginning the exercise.
- Emphasize to students the importance of reporting accidental spills and exposures.

Document Practices

- Require students to sign safety agreements explaining that they have been informed about safety precautions and the hazardous nature of the organisms they will handle throughout the course.
- Maintain student-signed safety agreements at the institution.

- Prepare, maintain, and post proper signage.
- Document all injuries and spills; follow school/college/university policy, if available.
- Make Material Safety Data Sheets (MSDS) available at all times; follow institutional documentation guidelines regarding number of copies, availability via print or electronic form, etc.
- Post emergency procedures and updated contact information in the laboratory.
- Maintain and make available (e.g., in a syllabus, in a laboratory manual, or online) to all students a list of all cultures (and their sources) used in the course.

—Text from Table 1, Emmert et al. (2013)

BIBLIOGRAPHY

Anderson C. 23 June 2008. The end of theory: the data deluge makes the scientific method obsolete. *Wired.* https://www.wired.com/2008/06/pb-theory/.

Anonymous. 2016. Go forth and replicate! *Nature* **536:**373. https://www.nature.com/news/go-forth-and-replicate-1.20473.

Barrowman N. Summer/Fall 2014. Correlation, causation, and confusion. *The New Atlantis*, No 43, 23–44. http://www.thenewatlantis.com/publications/correlation-causation-and-confusion.

Cobb M. 2014. Oswald Avery, DNA, and the transformation of biology. *Curr Biol* **24:** R55–R60. http://dx.doi.org/10.1016/j.cub.2013.11.060.

Emmert E. and the ASM Task Committee on Laboratory Biosafety 2013. Biosafety Guidelines for Handling Microorganisms in the Teaching Laboratory: Development and Rationale. *Journal of Microbiology & Biology Education, May 2013, p. 78–83.* http://dx.doi.org/10.1128/jmbe.v14i1.531http://dx.doi.org/10.1016/j.cub.2013.11.060.

Regalado A. 2015 April 15. CRISPR patent fight now a winner-take-all match. *MIT Technol Rev.* https://www.technologyreview.com/s/536736/crispr-patent-fight-now-a-winner-take-all-match/.

Sarkar RR. 2016. The big data deluge in biology: challenges and solutions. *J Informatics Data Mining* **1:**14. http://datamining.imedpub.com/the-big-data-deluge-in-biology-challenges-and-solutions.php?aid=9724.

Identifying the bacteria causing infections in hospital patients

Debilitated patients are getting infections while in the hospital. The clinical microbiology laboratory in the hospital has isolated and identified the microbe causing the illness. They suspect that the infections are due to contamination of hospital items such as catheters and intravenous tubing, and are asking you as independent microbiologists to test this idea. Using sterile swabs, the laboratory took samples at various locations in the hospital, both where illness was observed and where it was absent. Bacteria on each swab were then transferred to agar medium in a petri dish and allowed to grow before the dishes were given to you. To minimize bias, the identity of the isolated strain will only be revealed after you have completed your testing.

QUESTIONS BEFORE YOU BEGIN THE CHALLENGE

1. State the hypothesis being tested in this challenge.
2. What would be the expected results if the hypothesis is correct?
3. Would these results *prove* that the hypothesis is correct? If not, can you think of an alternative explanation?
4. What would be the expected results if the hypothesis is incorrect?

Strategy for Challenge One

1. *Streak the growth from a swab plate for isolated colonies.*
2. *Select different isolated colonies as pure cultures.*
3. *Characterize the microorganisms from each pure culture.*
4. *Identify the different bacteria at each site.*
5. *Determine which sites are contaminated with the microorganism causing disease.*
6. *Determine if these are the sites where disease is occurring.*
7. *Decide whether hospital items are the likely source of the disease-causing bacteria.*

Lab One

BACKGROUND

Diversity and pure cultures

KEY POINTS

- There are a very large number of bacteria on the planet, and these bacteria are extraordinarily diverse.
- Microbiologists estimate bacterial diversity from the diversity of DNA sequences, a procedure known as metagenomics.
- The different bacteria at a particular site are collectively referred to as the microbiome.
- Pure cultures are essential for studying the properties of a single microorganism.
- Pure cultures can be obtained by depositing a single cell on agar medium. The colony formed after successive cell divisions is a pure culture.

We live on a planet where life is overwhelmingly microbial. The number of human beings on Earth is approximately 7 billion (7×10^9), a large number, but insignificant compared to the estimated 10^{30} bacteria that share our world (Kallmeyer et al., 2012; Whitman et al., 1998). Of course, different bacteria are not all present in equal numbers. Probably the most abundant bacterium is *Pelagibacter ubique*: there are about 2×10^{28} cells in our oceans, making it (so far at least) the most common organism on the planet. Oddly, it wasn't discovered until 2002 (Morris et al., 2002).

Bacteria are not only very abundant but also extraordinarily diverse. There are about 8.7 million species of eukaryotes (plants, animals, and single-celled organisms included) (Mora et al., 2011). It is difficult to estimate the number of bacterial species—indeed, how a bacterial species should be defined is still being debated. One rough estimate is that there could be as many as 10^9 different types of bacteria on Earth (Dykhuizen, 2005). How can we guess that there is such a large number, particularly when only about 11,000 of these have been isolated and properly characterized (Whitman et al., 2015)? Scientists select different sites (forest soil, sandy soil, riverbeds, surface of the skin, and so on), sample the microbial DNA at each location, and then estimate bacterial diversity from the number of different base sequences. In other words, the presence of a unique bacterial type is inferred from the distinctly different base sequence of its DNA. An approach of this kind is called **metagenomics**, and each different type of bacteria discovered in this way is referred to as an **operational taxonomic unit** (OTU). This term sidesteps the issue of how a bacterial species should be defined and emphasizes that identification has

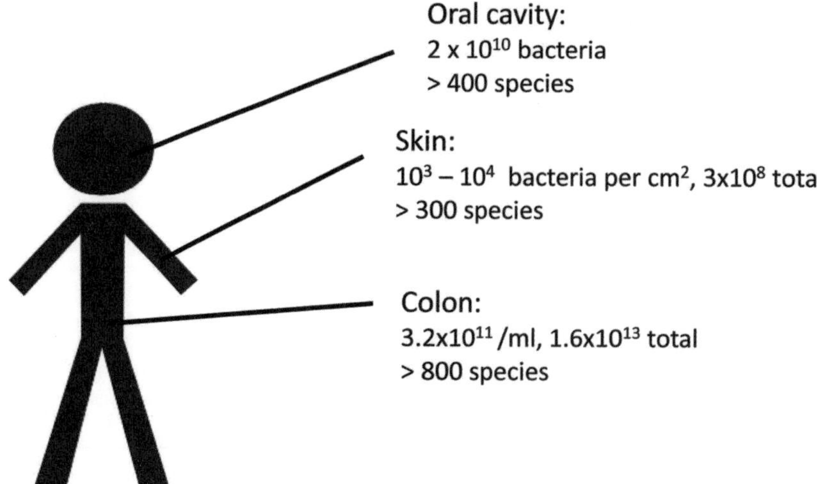

Figure 1-1 Number and diversity of bacteria associated with the human body. Source of data: NIH Human Microbiome Project. Stick figure image from Clipartfest.com Human figure from clip-art created by Nicu Buculei.

been based solely on DNA sequence. The **microbiome** is the set of all the different OTUs at a particular site. Metagenomics has revealed that even a small difference in variables such as temperature, pH, salinity, or oxygen content has an impact on the prevalence and kinds of bacteria in the microbiome (Griffiths et al., 1996).

Of course, we care most about the bacteria we come in contact with and that affect our lives. Bacteria have formed a relationship with many multicellular organisms, and humans are no exception. Figure 1-1 shows the estimated number and diversity of bacteria at various locations on our body. Altogether the bacterial cells outnumber our own cells, and, although bacterial cells are small, we are carrying around at least 2 to 6 pounds of them, made up of more than 10,000 different species (or, more properly, OTUs) (National Institutes of Health, 2012).

The fact that there are usually many different bacteria at a single site has been known since the dawn of microbiology, even though the extent of this diversity was not. Mixed populations create at least two problems in studying microorganisms. The first is lack of reproducibility: every time a site is sampled, the number and diversity of the microorganisms in the sample are likely to be different. You can characterize the skin bacteria from a volunteer one week, then two weeks later go back to the same site on the same volunteer and get a different result. The second is the problem of determining cause and effect. Bacteria are responsible for many of the properties of our bodies and our environment, but with so many different kinds of bacteria at the same location, how do we determine which are responsible for what? For example, plants require nitrogen, and much of it is derived from the decomposition of organic matter by bacteria and fungi. During decomposition, the nitrogen in organic molecules is mostly converted to ammonium ion (NH_4^+), which is not readily accessible to plants. In the soil, other bacteria convert NH_4^+ to nitrates, which are easily absorbed by plants. The soil contains hundreds or thousands of different species: which ones carry out this very important conversion?

To avoid these problems, microbiologists rely on **pure cultures**. A pure culture contains only one kind of bacteria so there is no variation in sample composition and cause and effect can be more readily determined. The importance of pure cultures is embodied in **Koch's postulates**, a cornerstone of medicine for over 100 years. Koch's postulates are a set of criteria for determining whether a particular microorganism causes a disease. A central tenet of the postulates is that the microorganism must come from a sick animal, must be grown in pure culture on artificial medium, must be shown to cause the disease when subsequently introduced into a healthy animal, and finally must be recultured from the sick animal. We now recognize that there are limitations in applying Koch's postulates (a disease can be caused by the combined effect of several microorganisms), but obtaining pure cultures remains an important part of microbiology.

If pure cultures are essential for determining cause and effect, how are they obtained? Most bacteria grow by cell division, one cell becoming two, and then these dividing again to become four cells, and so on. Therefore, if a culture is started from one cell, it must be a pure culture, a clone of genetically identical cells. Microbiologists often use solid agar medium in a petri dish to obtain pure colonies. A small number of cells are deposited on the surface, each well isolated from the other (you will see in the lab how this can be done easily). As each cell grows and divides, a colony of cells is formed on the agar medium. Because each colony originated from a single cell, all the cells are genetic clones of the original. Therefore, it is a pure culture.

Bright-field and phase-contrast microscopy

KEY POINTS

- Most bacterial cells can be visualized with a light microscope.
- The maximum useful magnification is 1,000X.
- In traditional bright-field microscopy, bacterial cells are usually stained to increase their contrast.
- Staining usually kills or alters the properties of bacterial cells.
- Phase-contrast microscopy overcomes the problem of staining by having an optical system that increases the contrast of unstained specimens.
- Phase-contrast microscopy is the best method for visualizing living bacterial cells.

Can a typical bacterial cell be viewed by the light microscope? Bacterial cells commonly have a diameter of about 0.5 to 1 μm and a length of 1 to 4 μm ($1\ \mu m = 10^{-6}$ meters). The maximum magnification of most laboratory light microscopes is 1,000X, so a 1-by-2-μm bacterial cell would appear 1×2 mm at this magnification, a size the human eye can detect. However, whether we can actually see such an object depends also on the **resolving power** of the microscope at this magnification. Suppose you are looking at two small objects under the microscope. Initially the two objects are far apart and there is no problem distinguishing each of them. Now the two objects are moved closer together: at some point, they will appear to merge into one blurry object. The higher the resolving power, the more closely two objects can be without their becoming fuzzy or merging into one. Put another way, suppose the word "microbiology" was written in very small letters so that it could only be read by using a microscope. You select a magnification so that the letters will appear as large as the ones on this page, large enough for reading. If the letters can be resolved by the microscope, you might see something like this: **microbiology**. However, if the resolution is inadequate, then you might see **microbiology**. The magnification and letters are the same, but the image is readable only if there is sufficient resolution to keep the letters separate.

Resolution depends on the wavelength of light. It turns out that the maximum achievable resolution at 1,000X is about 0.2 µm. In practice, this means that particles this size and smaller would not appear as clearly defined objects under the microscope. Thus, most bacterial cells can be visualized by light microscopy. Light microscopes are able to achieve higher magnification, but the resolving power is independent of magnification. As a result, the image would appear larger but would look blurry with loss of detail. 1,000X is near the maximum usable magnification (i.e., no blurring) for light microscopes, which is why it is commonly the highest magnification for a light microscope. Recently, ultrasmall bacteria were identified: round cells with a diameter of approximately 0.2 µm (Luef et al., 2015). These organisms would be very difficult to see by ordinary light microscopy.

There is a significant problem with bright-field (ordinary) light microscopy when viewing bacterial cells and many other biological specimens. In **bright-field microscopy**, objects are visualized because they absorb light, making them appear darker than the background. Most bacterial cells absorb very little light, so that individual cells are nearly transparent and may be difficult to distinguish clearly (Fig. 1-2A and C). Over

Figure 1-2 Bacterial cells visualized at **(A)** 400X, no phase contrast, **(B)** 400X with phase contrast and **(C)** 1000X, no phase contrast.

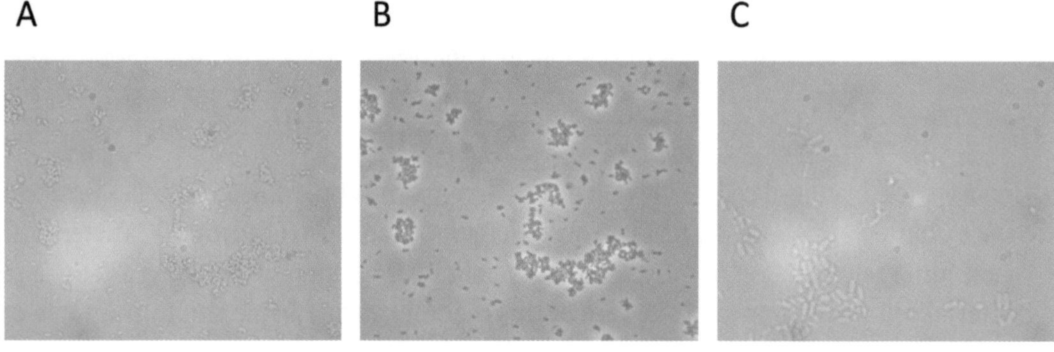

the years, many different stains were developed to solve this problem. Most stains bind to the bacterial cell but not to the background, although with negative staining it is the other way around. Some stains are useful for characterizing the cell. You will use one of the most important of these, the Gram stain, in the second lab.

The problem with classical staining is that during the process the cells are killed, or their environment significantly altered, making it difficult or impossible to observe live-cell behaviors such as motility. In addition, the stain can alter the structure or organization of cells in ways that can be misleading.

Phase-contrast microscopy is an important tool in microbiology because it achieves contrast without staining and therefore without killing the cell (Fig. 1-2B). In 1953, Frits Zernike won the Nobel Prize in physics for developing this technique. Zernike realized that light travels at a slower speed through a specimen than through the surroundings: this means that the phase of the light wave through the specimen is shifted. He developed an optical system so that this phase change was exaggerated and then recombined with surrounding light. The result was phase interference, the amplitudes of the light waves from the two sources canceling each other out and, therefore, the specimen appearing darker than the background. The nice thing about this is that living material can be observed directly. In addition, because organelles in a eukaryotic cell retard light to slightly different degrees than the cytoplasm, the amounts of interference and thus darkening are different. This allows these structures to be directly visualized as well. For the same reason, phase-contrast microscopy can reveal nucleoids in living bacterial cells, thus proving the existence of these subcellular structures (see http://schaechter.asmblog.org/schaechter/2013/04/pictures-considered-3-how-do-you-know-there-is-a-nucleoid.html). It is important to realize, however, that phase-contrast microscopy is still light microscopy and subject to the same limitations in resolution.

Lab One

1. ***Streak the growth from a swab plate for isolated colonies.***
2. *Select different isolated colonies as pure cultures.*
3. *Characterize the microorganisms from each pure culture.*
4. *Identify the different bacteria at each site.*
5. *Determine which sites are contaminated with the microorganism causing disease.*
6. *Determine if these are the sites where disease is occurring.*
7. *Decide whether hospital items are the likely source of the disease-causing bacteria.*

Learning Outcomes

After this lab, students will be able to:

- **a.** Make a wet mount.
- **b.** Transfer bacteria from solid media to solid media.
- **c.** Do a four-phase streak for isolated colonies.
- **d.** Identify the parts of a bright-field microscope.
- **e.** Examine cells using phase contrast or oil immersion.

I. Isolate bacteria from a mixed culture

Sampling from a culture grown in liquid or on agar medium inevitably involves picking up a very large number of cells. Even an invisible amount is likely to contain thousands of cells or more. If these were simply spread on agar medium, there would be so many cells that the plate would not contain the isolated colonies needed for a pure culture. For this reason, a sample must be diluted before it is placed on agar medium. One approach is simply to make different dilutions of a mixed culture by resuspending the cells in sterile liquid, then spreading a sample of each dilution on a separate plate (petri dish) containing agar medium. At the right dilution, only a few cells will be deposited on the medium and each of the resulting colonies will be a pure culture. This can be time-consuming, especially if you have no idea what the best dilution would be. A much faster method to make dilutions is by streaking with a wire loop, which you will learn during this lab. Essentially, by streaking you are using a sterile loop to dilute the sample progressively, so that eventually single cells are distributed onto a section of the agar medium. After incubation, these single cells form isolated colonies of pure cultures.

The class will be divided into groups, each receiving a different hospital sample. It will be the task of each group to identify the bacteria in their sample. Remember that samples are from both affected and unaffected areas of the hospital. However, you have not been told which samples belong to each of these groups, nor have you been told the identity of the pathogen isolated in the clinical microbiology lab.

The samples taken from the hospital sites undoubtedly contain more than one bacterial species. You will therefore need to isolate as pure cultures the different bacteria in the samples before each can be characterized.

PROCEDURE

Streaking for isolated colonies

Each member of your group should obtain a tryptic soy agar (TSA) plate and label it completely and legibly on the back with a marking pen. Include your name, date, and group number if one is assigned. Do not label plates on their covers. Write near the edge of the plate, as writing across the middle can obscure your view of isolated colonies.

1. Sterilize your loop using a Bunsen burner (or preferably an electric incinerator) and let it cool for 5 to 10 seconds. Do not heat the chuck or loop handle. Your instructor may provide instead sterile, disposable loops. See **Technique Box 1** for tips on flame sterilization.

SAFETY Never leave an open flame unattended. Always make sure that the flame is out and you have completely turned off the gas supply before you leave the lab.

2. Touch the cooled loop to the bacteria on the swabbed plate and transfer the sample to a position near the periphery of the new plate. Do not transfer a large amount of culture: your sample should be barely visible.
3. Gently streak back and forth from the periphery. Extend the streaks about a quarter of the circumference of the plate. As you streak, move the loop about one-third of the way toward the center of the plate (Fig. 1-3). Try not to gouge the agar.
4. Rotate the plate about 90°.
5. Sterilize the loop as in step 1 (or use a new disposable loop). Do not resample from the swabbed plate.
6. Touch the loop to one end of phase 1.
7. Streak back and forth, again extending the streak about one-fourth of the plate circumference and drawing the loop one-third of the way toward the center. This is phase 2. Pass through the first phase a few times at first, but then as you continue streaking, do not continue contacting this phase.
8. Repeat steps 5 to 8 for phases 3 and 4. Remember to sterilize the loop each time and to contact the previous phase only a few times during the beginning of the streak. When you streak for phase 4, distribute the cells into the center of the plate, being careful not to contact any of the other phases.

Place the plates in an incubator set at 37°C. Unless stated otherwise, plates should always be incubated inverted, with the cover on the bottom, so that drops of moisture do not fall onto the surface of the medium. Remove the plates (or your instructor will remove them) after approximately 24 hours and store them at 4°C. In the first phase, a large number of cells were deposited, and these will probably form a continuous patch of cells rather than isolated colonies. In that case, the growth is said to be confluent. By the fourth phase of streaking you should observe isolated colonies. Successful and unsuccessful streak plates are shown in Fig. 1-4. In Fig. 1-5, a mixed culture of *Staphylococcus epidermidis* and *Serratia marcescens* was streaked onto agar

TECHNIQUE BOX 1

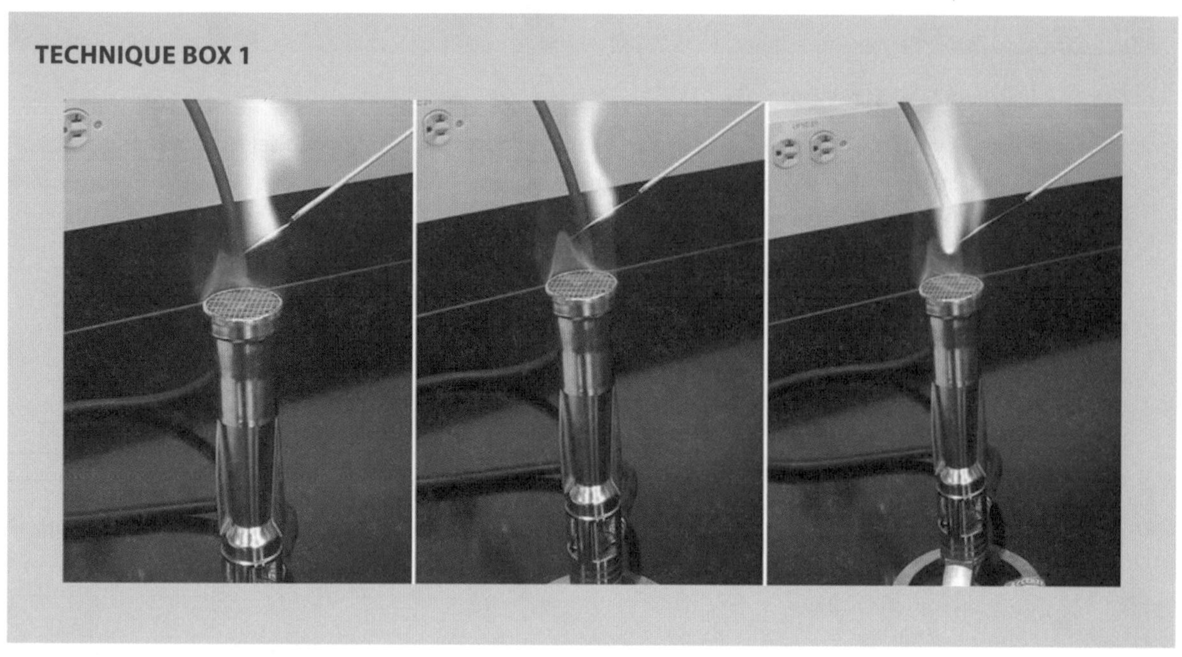

Figure 1-3 Technique for streaking bacteria on agar medium.

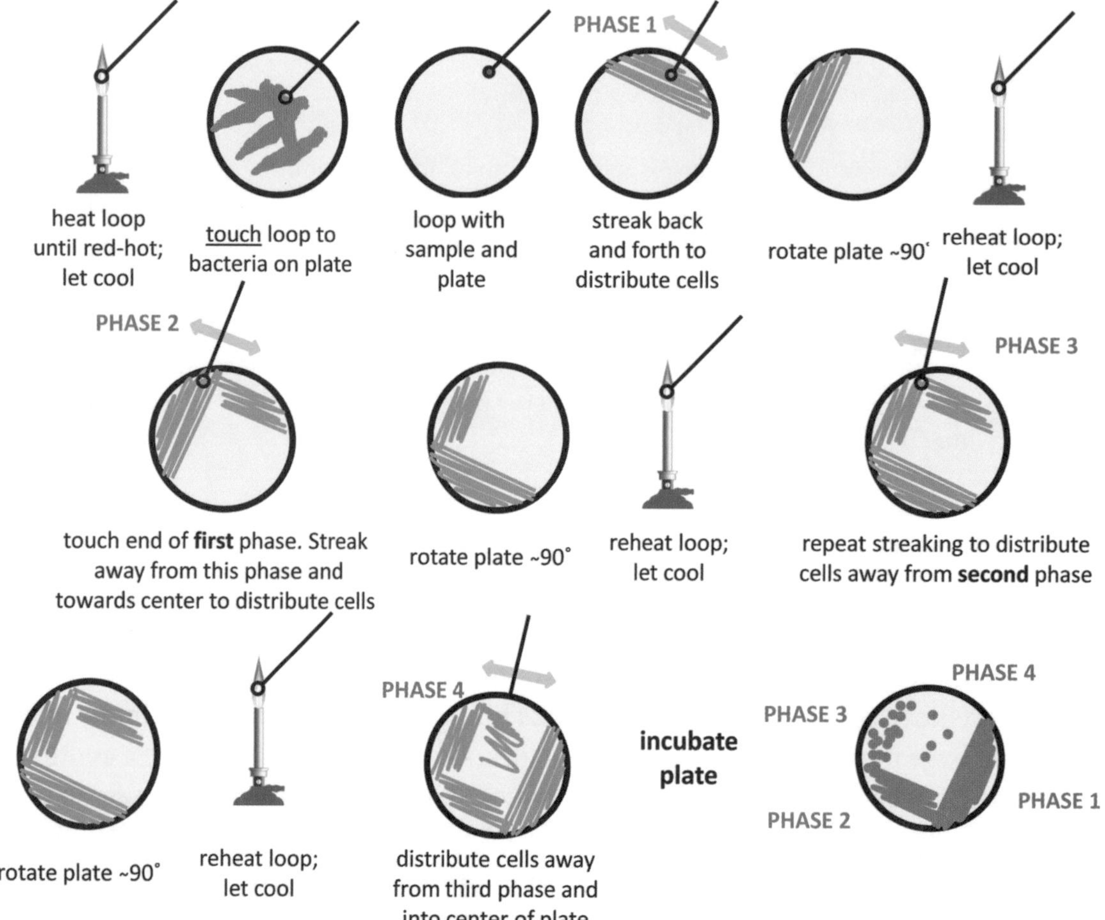
heat loop until red-hot; let cool
touch loop to bacteria on plate
loop with sample and plate
PHASE 1
streak back and forth to distribute cells
rotate plate ~90˚
reheat loop; let cool
PHASE 2
touch end of **first** phase. Streak away from this phase and towards center to distribute cells
rotate plate ~90˚
reheat loop; let cool
PHASE 3
repeat streaking to distribute cells away from **second** phase
rotate plate ~90˚
reheat loop; let cool
PHASE 4
distribute cells away from third phase and into center of plate
incubate plate
PHASE 4
PHASE 3
PHASE 2
PHASE 1

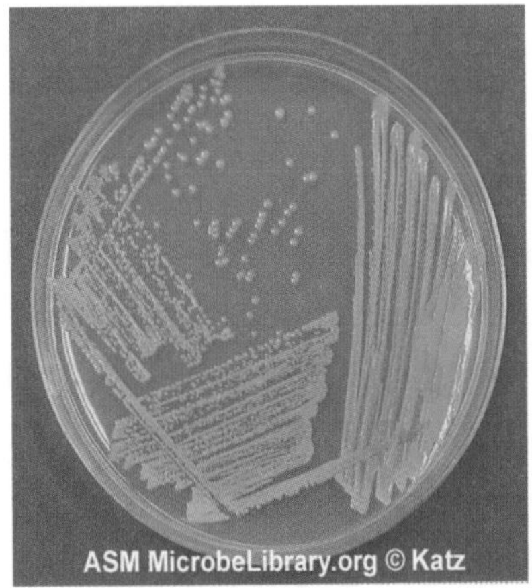

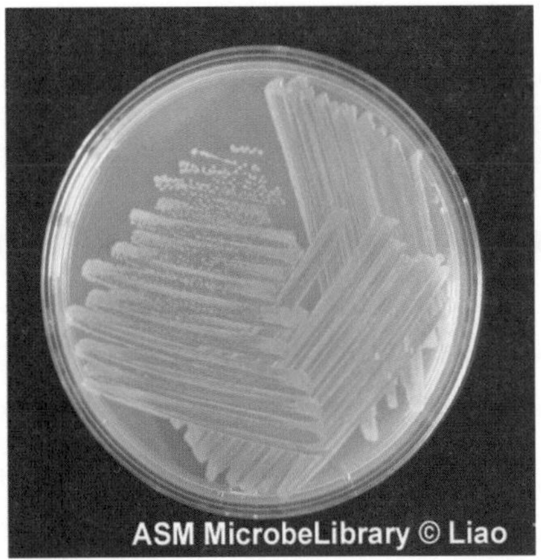

Figure 1-4 A successful (left) and unsuccessful (right) streaked plate. ASM MicrobeLibrary. Credit: (left) D. Sue Katz, Rogers State University (right) Min-Ken Liao, Furman University.

medium. The smaller colonies are *S. epidermidis* and the larger *S. marcescens*. There are two things to notice about this. First, bacteria often have distinct colony morphologies, and this can aid in their identification. Second, the size of the colony is a rough indication of the rate of growth (how often the cells have divided) to form the colony.

Poor technique that results in no isolated colonies is generally due to one or more of the following:

1. You are initially applying too much of the source bacteria or you are carrying too many cells from one phase to the next. As a result, the cells are never diluted enough during streaking to result in isolated colonies.
2. You are mistakenly resampling the original plate culture before streaking each phase.
3. While streaking toward the center of the plate, you are touching other phases.
4. You are forgetting to sterilize the loop before streaking each phase.

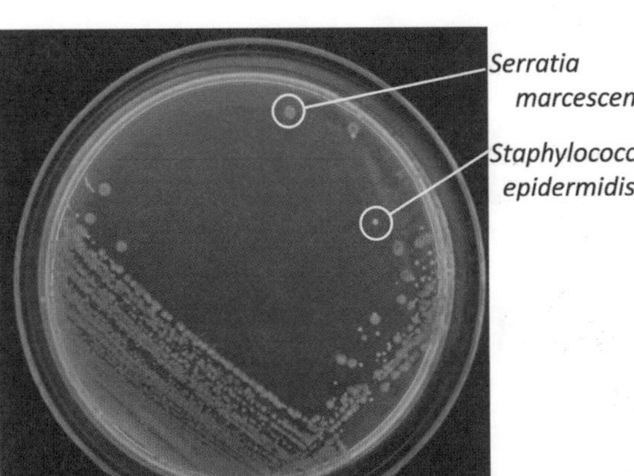

Figure 1-5 Mixed culture of *Serratia marcescens* and *Staphylococcus epidermidis* streaked on a TSA plate. Each isolated colony represents a pure culture of one of the two microorganisms. Notice the two colony sizes. Credit: Seth James.

5. You are not letting the loop cool down sufficiently before streaking. You can hasten cooling by touching the sterile loop to a sterile location on the agar medium.
6. Plates are wet. Examine your plate for surface moisture before beginning.

II. Examine bacterial cells under the microscope

Figure 1-6 is a photograph of a modern microscope. While microscopes differ in details, all have the same basic parts. Before using the microscope, familiarize yourself with:

1. The light source. This will have an on/off switch and a dial to regulate light intensity. Some simple microscopes instead have a mirror to reflect light from an external source.
2. The light condenser, which focuses the light onto the specimen.
3. The stage, which supports the glass slide having the specimen. The stage has an opening that allows light to pass through.
4. The specimen holder, which keeps the slide steady and limits accidental movement of the slide while viewing. Often the stage and/or specimen holder is horizontally adjustable so that different parts of the specimen can be easily viewed.
5. The objective lenses and revolving turret. The lens holders are mounted on a wheel that allows you to swivel from one lens to another. Always use the turret when changing lenses. If you grab a lens holder, it could put the lens out of alignment.
6. The coarse and fine focus adjustment knobs. These change the distance between the objective lens you are using and the stage. Never attempt to focus

Figure 1-6 Major parts of a modern microscope. Photograph courtesy AmScope.

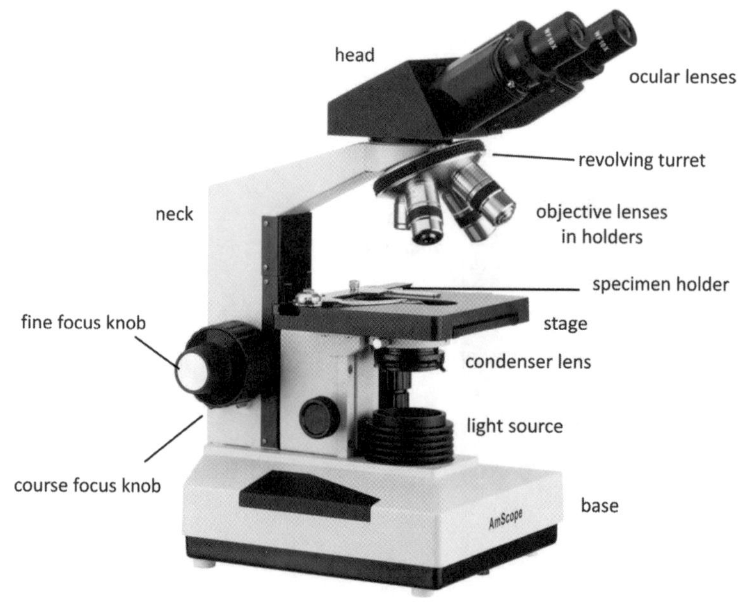

by bringing the stage and lens closer together, because some lenses can be brought into contact with the specimen, and this can scratch the lens. More detailed guidance on focusing is given below.

For a microscope like the one in Fig. 1-6, the ocular and objective lenses together magnify the specimen. Instruments with more than one magnifying lens are called **compound microscopes**, and they achieve greater magnifications than are possible with a single lens. In a compound microscope, the overall magnification is the product of the magnifying power of the objective lens (the one close to the specimen) and the ocular lens (the one close to the eye). Typically, ocular lenses are 10X, in which case the final magnification for 40X and 100X lenses are 400X and 1,000X, respectively.

Microscopes look sturdy, but in fact they are delicate instruments. Rough usage can result in the light path no longer being in alignment. If you need to move the microscope to a new location, always use two hands, one supporting the base and the other holding the neck (these parts are shown in Fig. 1-6).

PROCEDURE

Making a wet mount

1. Use a sterile loop or eye dropper to transfer a small drop of water onto a microscope slide.
2. Sterilize your loop by placing it in a Bunsen burner or electric incinerator until it is red-hot. Let cool.
3. Touch your loop to your swab sample so that a nearly invisible amount is taken up, then transfer these cells to the drop. During the transfer, use a circular motion with the loop so that the cells are uniformly suspended in the drop. If you use too many cells, you will be looking at a paste of bacteria on the slide and will have difficulty seeing individual cells.
4. Add a coverslip and examine by bright-field or phase-contrast microscopy (40X objective lens). Trapped air bubbles are a common occurrence and can be useful (below). The bubbles can be minimized by holding the coverslip at an angle to the slide, then slowly placing it over the sample.

PROCEDURE

Using the microscope

Your instructor will demonstrate the instrument you will be using. Bacterial cells are very small and might be difficult to see at first. Many compound microscopes are **parfocal**, meaning that if one objective lens is in focus the others will be in focus or nearly so. It is easier to focus with lower magnifications, for example, with a 10X objective lens in place.

1. Using the lowest magnification objective (usually 4X), focus on the coverslip edge. You will probably need both the coarse and fine adjustments for this. When the edge is in focus, gently swing the objective lens with the next-highest magnification (usually 10X) into the light path (remember to use the turret to do this). If you focused correctly, the new lens should not strike the coverslip, but watch carefully to be sure. Notice that *the higher the magnification of the objective lens, the closer it is to the coverslip*. Be careful also not to change the position of the slide. You should be in nearly correct focus if the microscope is parfocal. You might not see the edge of the coverslip because the field of view is smaller at higher magnifications, but you should find it again by moving the slide slightly.
2. The next objective lens is usually 40X. Swing this objective into the light path, watching carefully that the objective does not touch the slide. If you are using phase contrast, make the necessary adjustments to the microscope. Again, focus on the edge of the coverslip.
3. Now move the slide so that the liquid sample under the coverslip is in the light path. The bacteria will probably not be in focus because the coverslip is much thicker than the cells, so you will still need to use the fine focus to see the bacteria. Never use the coarse adjustment at higher magnifications. If air bubbles are in the wet mount, focusing first on the edge of one of these will be helpful. Remember, though, that you want to look at cells suspended in liquid, not the interior of the air bubble.
4. If you used phase contrast, return the microscope to the original, bright-field settings. Turn the turret so that the 40X and 100X objectives are on either side of the light path. Add a drop of microscope oil to the coverslip, then move the microscope turret so that the 100X objective lens is in use. Again, watch carefully. The 100X objective should come into contact with the oil but should not hit the coverslip. If you feel any resistance, seek help from your instructor. Be careful not to put oil on any of the other objective lenses. You will probably need to refocus: remember to use only the fine adjustment. Under this magnification, notice if the cells have different sizes and shapes. Cell size and shape is another way bacteria are characterized. After you are finished, clean the 100X oil immersion lens with microscope cleaner and lens paper.

You should see individual cells under the microscope. If your microscope is fitted for phase contrast at this magnification, or if you have a phase-contrast microscope in the class, notice how this feature improves contrast and visibility.

You also might observe that some of the cells are moving. There are several possible reasons for this. If the cells appear to be twitching or moving a short distance in a random direction, it is probably due to Brownian motion, the random bombardment of molecules on the cell surface. If the cells are all flowing the same way across the field of vision, this is probably the result of convection caused by unequal heating of the sample. In neither of these cases are the cells truly motile. Motile cells will move faster and longer in a single direction, which will periodically change suddenly and randomly, usually every couple of seconds. Many bacteria, including many pathogens, are motile, and this is one of the ways bacteria are characterized. You will have a chance later to determine if the cells in your isolate are motile.

You might be wondering why oil is being used with the 100X objective lens. Light passing through the specimen is scattered (refracted) at the glass coverslip-air boundary so that some of this light does not pass through the objective lens. The oil reduces this scattering because it has nearly the same refractive index as glass. Thus, the lens is able to gather more of the light passing through the specimen. This broadens the field of view and improves the resolution.

❖ Preparation for next lab

This week each of you purified the bacteria in your hospital sample by streaking on agar medium. These plates were incubated at 37°C and then stored at 4°C. The experiments next week require fresh cells. Return to the lab approximately 24 to 48 hours before your class period and retrieve your plates from the 4°C refrigerator. Examine the streak plates from each member of your group and estimate the number of different colony types that were isolated from the swab sample. Each of you should select a well-isolated colony for further characterization. Make sure your group has selected a representative of all the different colony types, even if you are uncertain whether the colonies are different or not. Using a permanent marker, draw a circle on the bottom of the plate around the colony you selected: this is your primary isolate. Write your initials next to the colony. Obtain a new TSA plate and make a four-phase streak plate using your selected isolate. Incubate at 37°C for your next scheduled lab.

QUESTIONS

Questions are designated B1 to B6 according to the six levels of Bloom's taxonomy.

1. Define a pure culture. Why is it important for microbiologists to work with pure cultures? (B1)
2. Wet plates should not be used for streaking. What would be the result if you used a wet plate? (B2)
3. Labeling petri plates completely is important for identification. Why are plates labeled on the bottom rather than the top of the dish? (B2)
4. In the procedure for streaking on agar medium, why is it important to sterilize your loop before each new phase? What would happen if you forgot to do this? (B2)
5. The unknown bacteria are being incubated at 37°C, which is the normal body temperature of humans. Why do you think this temperature was chosen? (B2)
6. Certain pathogenic *Escherichia coli* can infect the kidney. Would you expect these cells to be motile? (B3)
7. *Acinetobacter baumannii* is an emerging pathogen that can be found in hospitals not only on tubing and other items used for treatment but also on door handles, curtains, furniture, cleaning supplies, and so on. One of the characteristics of this organism is that it is

very resistant to desiccation (drying-out). Explain why this property is an important consideration in controlling infection. (B4)

8. Not all bacteria form colonies: *Proteus mirabilis*, for example, migrates along the surface of the agar medium (Fig. Q1-1). The migration in the picture is several days after the center of the plate was inoculated. The migration is periodic, resulting in the wave-like appearance. How might the presence of *Proteus* in a sample complicate the purification of different bacteria? (B4)

Figure Q1-1 *Proteus mirabilis* spreading over the surface of a plate instead of forming colonies. The plate was inoculated in the middle. During growth, the bacteria moved outward in a series of waves. Image from http://schaechter.asmblog.org/schaechter/2011/11/are-you-me-or-am-i-you.html.

9. At the end of this challenge, the class will compare the results for the different samples and conclude whether the isolate causing infection is probably coming from contaminated hospital items. If you all knew the identity of the isolate, and also which samples were from an area with infections, what kinds of biases could affect your analysis? (B5)

10. Two students streaked cells onto a plate for the first time. The results are shown below (Fig. Q1-2). Who did the best job? Describe the error(s) that were probably made in the other case. (B5)

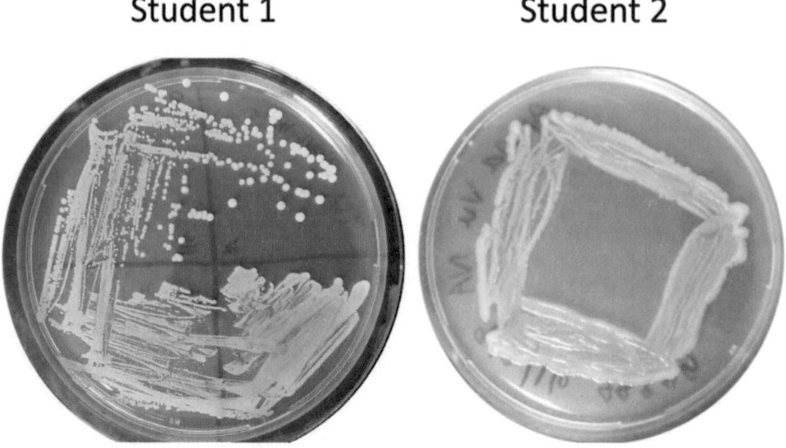

Figure Q1-2 Two attempts to streak on agar medium for isolated colonies.

11. Periodontitis is the major cause of tooth loss in the United States. The disease is caused by a group of microorganisms in the space between the gum and the tooth. How are Koch's postulates complicated by this disease? (B5)

Lab Two

BACKGROUND

Colony morphology and optimum temperature for growth

KEY POINTS

- Many bacterial species have a characteristic colony morphology that can be useful for identification.
- The appearance of a colony may be affected by conditions such as temperature and growth medium.
- Different bacteria grow in a wide range of temperatures, but those most commonly encountered are mesophiles, which grow best between room temperature (25°C) and normal body temperature (37°C).
- Within the temperature range of mesophiles, the optimum temperature for growth is characteristic of each species and aids in identification.

Often, a bacterial species will have a characteristic colony morphology that can be helpful in identifying the organism. For example, some of the organisms you are likely to encounter form the colonies shown in Fig. 1-7. However, it is important to know the conditions of growth during formation of the colony: in particular, the choice of temperature and growth medium (Fig. 1-8A and B) can affect its appearance.

Some features to look for are:

Shape, border, and elevation: Is the colony round, or does it have an irregular shape? Are the edges of the colony smooth or wavy? Does the colony appear round or flat?

Color: Some bacteria form colonies with characteristic colors. Examples include *S. marcescens* (Fig. 1-5), which forms pale pink colonies at 37°C and dark red colonies at 30°C (Fig. 1-8B). Also look for pigments diffusing from the colony; some, like those produced by *Pseudomonas aeruginosa*, are fluorescent under UV irradiation.

Texture: Does the colony have a smooth or irregular surface? Does the colony appear dry or wet? Some of the terminology used to describe the characteristics of colonies is given in Fig. 1-9.

Since bacteria grow in a large number of different habitats, it is easy to understand why they might have different temperature optima for growth. Most of the bacteria

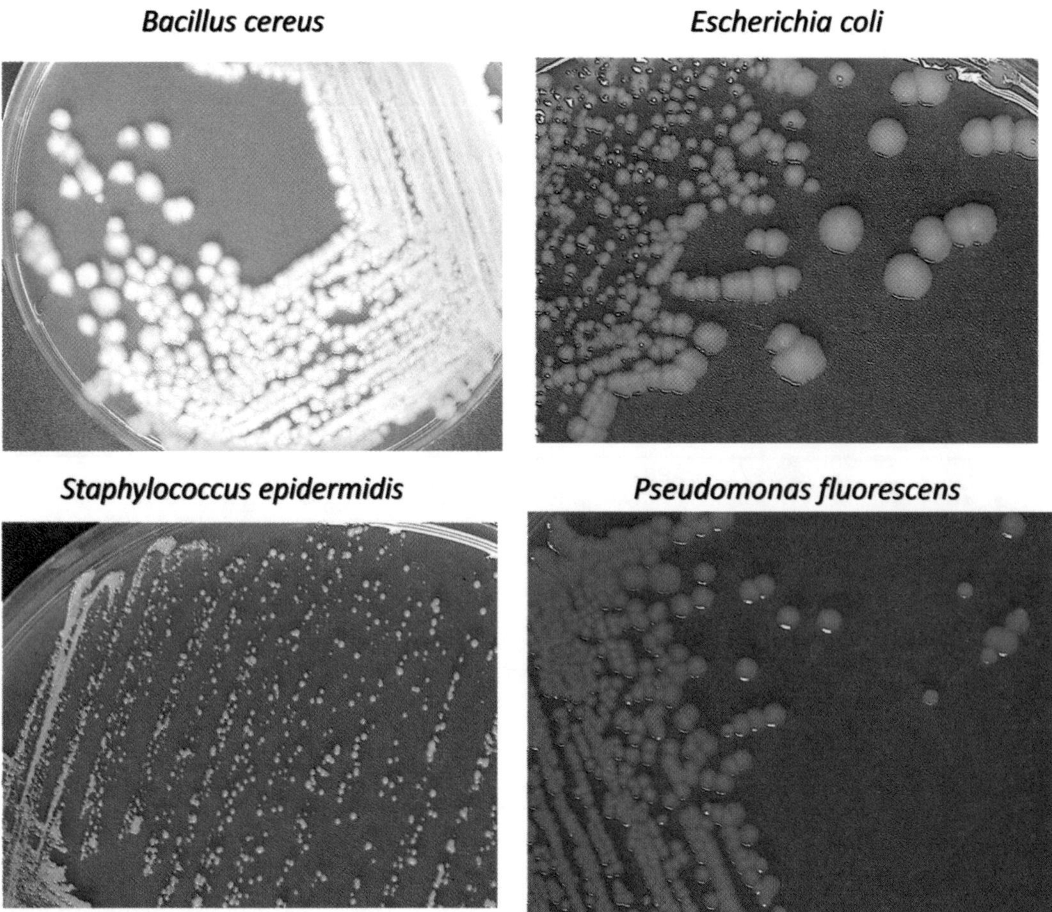

Figure 1-7 Colonies of different bacteria likely to be encountered in this course. *Bacillus cereus*: Dry-looking, flat colonies with irregular borders are characteristic of many species of Bacillus. Older colonies can become quite large. *Escherichia coli*: Medium-sized, light tan colonies have a glistening, raised surface and irregular margins. Laboratory strains may be drier in appearance. *Staphylococcus epidermidis*: Small white colonies, typically convex with regular borders. Colonies may have markedly different sizes. Colonies of many other staphylocci are similar. *Pseudomonas fluorescens*: Glistening, circular colonies with regular borders.

we regularly encounter are **mesophiles**, which grow between approximately 20 and 45°C. This includes not only common environmental temperatures but also normal body temperature (37°C). Many mesophiles become rapidly inactivated at temperatures above 45°C. This is the basis of pasteurization, which inhibits food spoilage by raising the temperature during processing. Usually, the temperature required for pasteurization is sufficiently low to preserve most of the qualities of the food. Of course, refrigeration is also used to preserve foods because mesophiles usually do not grow well at low temperatures.

It is important to distinguish between the temperatures required for growth and for survival. Some bacteria can last for a long time at temperatures that do not support growth. A well-known example is *Clostridium botulinum*, which produces the toxin responsible for botulism. This organism forms spores that can easily survive the

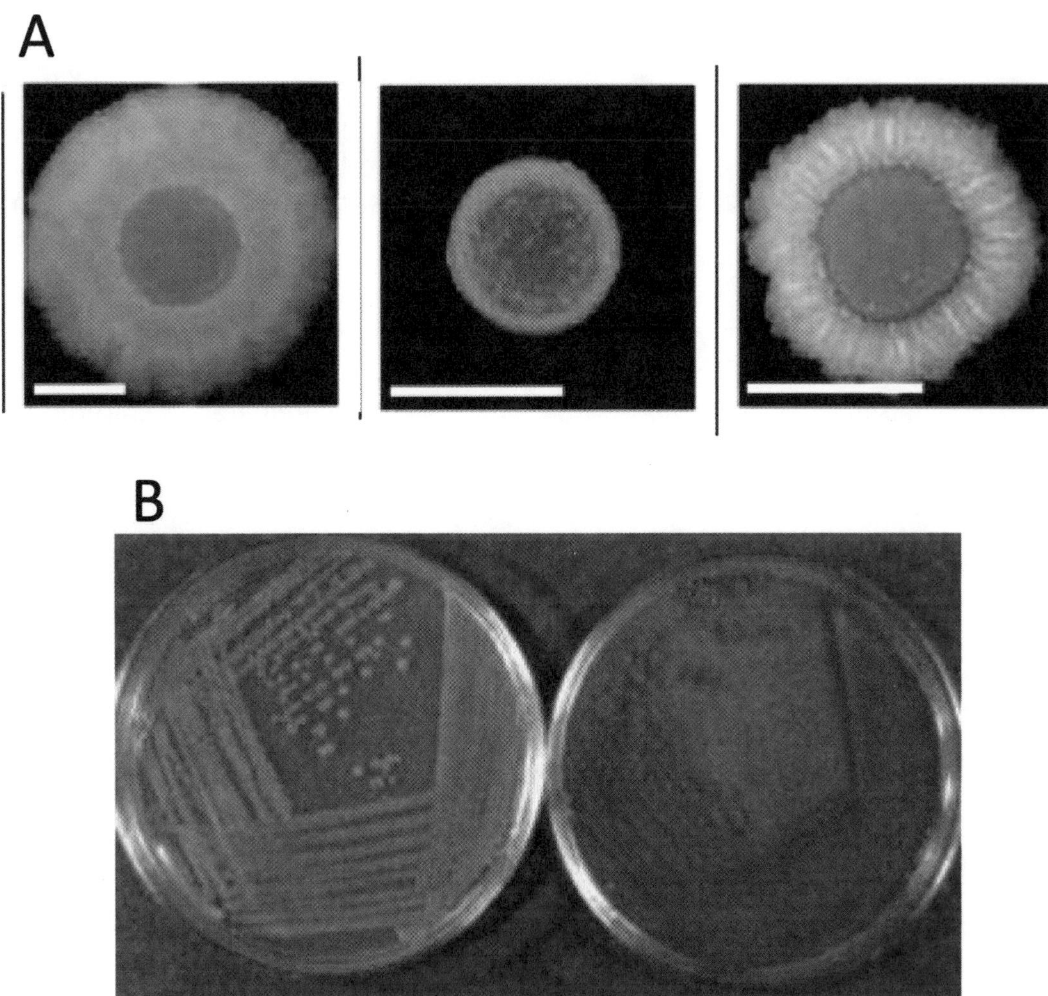

Figure 1-8 **(A)** Colony characteristics of the same strain of *Bacillus subtilis* grown on three different media. Modified from Lei *et al.* (2003) **(B)** Color of *Serratia marcescens* grown at 37°C (left) and 25°C (right). From Leboffe and Pierce (2012).

pasteurization process and many other harsh conditions. For this reason, the canning of most foods requires a high temperature and pressure (typically 121°C) to kill the spores. If these conditions are not reached, some of the spores can survive, germinate in the stored food, and produce the toxin. This is why home canners generally use a pressure cooker, a pot where air pressure is increased, therefore allowing steam to reach a temperature higher than 100°C.

Psychrophiles typically grow between –20 and 10°C and are found in cold environments, both in the oceans and on land. They have a number of interesting adaptations that allow them to survive in the cold (D'Amico et al., 2006). One important problem they face is that the activity of enzymes drops with temperature, which could result in an unacceptably low rate of catalysis for enzymes active in the mesophilic range. Many psychrophilic bacteria appear to solve this problem by having enzymes that are *less* stable than their counterparts in mesophilic bacteria. Because

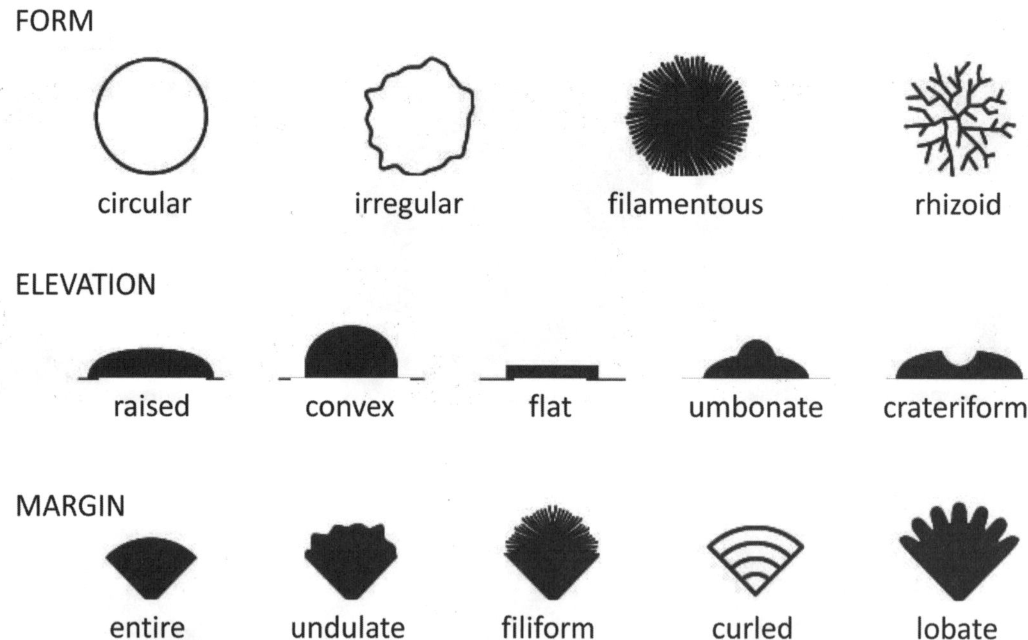

Figure 1-9 Diagram of colony types and terms used. Adapted from image published by Microbiology Society UK.

of their reduced stability, these enzymes retain sufficient flexibility at low temperatures to carry out an acceptable rate of catalysis.

Thermophiles are microorganisms that can grow at high temperatures, in some cases around the temperature of boiling water. They are commonly found in areas where there is volcanic activity, both on land and in the oceans. Many thermophiles are *Archaea*, which are not bacteria but a third domain of life distinct from both bacteria and eukaryotic organisms. However, there are bacterial thermophiles as well.

The extreme environments favored by many psychrophiles and thermophiles generally place them outside the interest of medical and food microbiologists. However, bacteria that are mesophiles, but able to grow outside of the usual mesophilic temperature zone, can cause problems. *Yersinia enterocolitica* is a cold-tolerant mesophile that can grow well at 4°C, within the temperature range of psychrophiles. This organism is a human pathogen that can invade the gastrointestinal tract, causing severe pain, diarrhea, and other symptoms of food poisoning. The symptoms sometimes resemble appendicitis and can result in unneeded surgeries. Because of its cold tolerance, refrigeration is not a safeguard against foods contaminated with *Y. enterocolitica*, although food poisoning by this organism is relatively uncommon. More seriously, however, *Y. enterocolitica* can grow in blood supplies stored at 4°C. This organism accounts for almost half the cases of posttransfusion sepsis (extreme inflammatory reaction to infection), with a mortality rate around 50% (Guinet et al., 2011). Fortunately, contamination of blood with *Y. enterocolitica* is very rare.

Even for mesophilic bacteria, knowing the temperature optimum for growth can contribute to the identification of a microorganism. For example, the pseudomonads are a large group of organisms mostly found in the soil and characterized by a diverse metabolism (some can degrade hydrocarbons) and the production of enzymes and toxins that are secreted into the environment. Many species are associated with plants and some can degrade plant tissue, causing disease. The optimum growth temperature is generally 25 to 30°C, but there is an important exception: *P. aeruginosa*. This organism, an important opportunistic pathogen of humans, grows well at the normal human body temperature of 37°C, and this no doubt contributes to its success in causing infections.

Cell shape and bacterial spores

KEY POINTS

- The cells of each bacterial species have a characteristic shape.
- Cells that are rod-shaped or nearly spherical are two common types, but cells can also be helical or branched, and other rare types exist as well.
- Species of *Bacillus* and *Clostridium* produce endospores, highly specialized cells that are able to withstand harsh conditions.

Bacterial cells have different shapes, and for some species, cell shape can be useful for identification. The two common cell types you are likely to encounter are:

Rod-shaped cells: Many bacteria are rod-shaped. *E. coli* cells (Fig. 1-10A) are typically about 0.25 to 0.5 µm in diameter and 1 to 2 µm in length. Many species are present as single cells, but a few form long multicellular filaments.

Round or oval cells: Cells that are nearly spherical or ovoid are called **cocci.** Different species of cocci often form characteristic arrangements of cells. Two medically important groups of bacteria are the staphylococci and streptococci (Fig. 1-10B and C). Staphylococci typically form grape-like clusters, whereas the streptococci form chains. The differences in the organization of coccoid cells are the result of the planes of cell division. Streptococci divide on only one plane, so chains of cells are formed. For staphylococci, the division plane is in three dimensions, selected sequentially. This results in the characteristic irregular cluster of cells.

Other cell types include:

Spiral-and comma-shaped cells: A surprising number of bacterial cells are spiral-shaped. Two major groups, the spirilla and the *Spirochaetes*, have this morphology, but they are in other ways very different. The spirilla are motile due to flagella at the ends of the cell (Fig. 1-10D). Many live in aqueous environments.

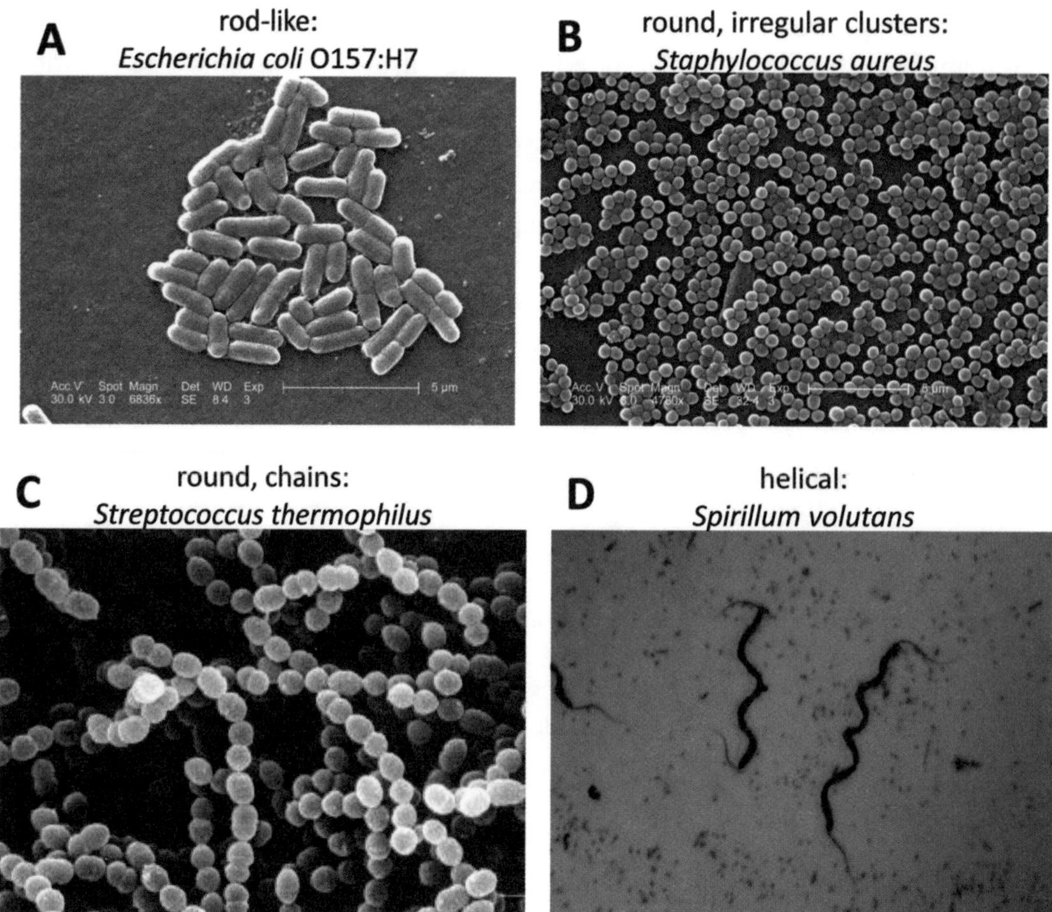

Figure 1-10 Shapes of bacterial cells **A.** Rod-like cells of *E. coli* strain O157:H7, a dangerous human pathogen. Some of the cells were in the process of dividing. Scanning electron micrograph. CDC Public Health Image Library, image #8800. Credit: CDC/Janice Haney Carr **B.** Round cells (cocci) irregular clusters of *Staphylococcus aureus*. CDC Public Health Image Library, image #7820 Credit: CDC/ Janice Haney Carr, Jeff Hageman. **C.** Round cells (cocci), chains of *Streptococcus thermophilus* Credit: [Eugenio Parente, Laboratory of Industrial Microbiology – Università degli Studi della Basilicata.] **D.** Helical cells (spirilla) of *Spirillum volutans* Credit: Steven R. Spilatro, Marietta College.

They include the important pathogen *Helicobacter pylori* (Fig. 1-11A), which causes ulcers and may be a contributing factor in stomach cancer. The *Spirochaetes* (Fig. 1-11B) have a flexible body that is part of their unique method of locomotion. They have axial filaments, which are much like flagella but are anchored at both ends of the cell. The filaments cause the cell to rotate, propelling it forward. The *Spirochaetes* do not make up a large group of bacteria but they nonetheless have a large impact: *Treponema pallidum* causes syphilis, and *Borrelia burgdorferi* Lyme disease.

Comma-shaped cells: are characteristic of the vibrios, including the infamous pathogen *Vibrio cholera*, which, as the name suggests, causes the deadly diarrheal disease cholera (Fig. 1-11C). The cells have a slight twist and may be considered abbreviated spirals. They are related to the spirilla.

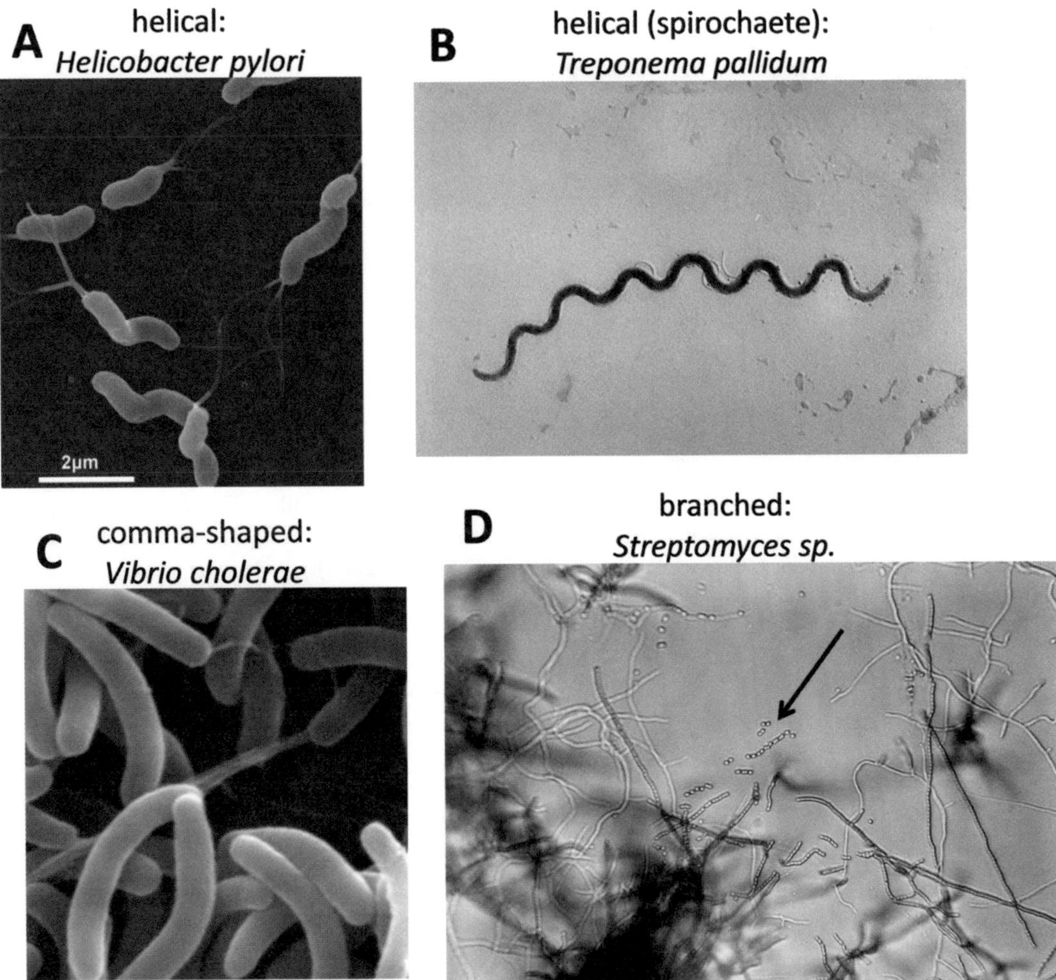

Figure 1-11 Shapes of bacterial cells (*continued*) **A.** Helical cells (spirilla) of *Helicobacter pylori*. Credit: Dr. Nina Salama, Fred Hutchison Cancer Research Center **B.** Helical cells (spirochaetes) *Treponema pallidum*, the causative agent of syphilis. CDC, Public Health Image Library, image #14969. Credit: CDC/ Susan Lindsley **C.** Comma-shaped cells of *Vibrio cholerae* Credit: Dartmouth Electron Microscope Facility, Louisa Howard. **D.** Branched cells of streptomyces (species unknown). Note the branching cells. The long chains of small cells (arrow) are spores. CDC, Public Health Image Library, image #2983 Credit: CDC/ David Berd.

Branched cells: Streptomycetes are a diverse family of bacteria found in the soil and known for their production of antibiotics. They are quite abundant and are responsible for the characteristic odor of soil. Streptomycetes superficially resemble fungi. They generally have long, branched cells (Fig. 1-11D).

Spores: Several important groups of bacteria, including some of medical interest, produce spores. Bacilli and clostridia produce **endospores** (Fig. 1-12A), which develop inside the cell. As mentioned above, the spores of these organisms confer protection from environmental insults such as high temperature. They have little or no metabolism and contain a desiccated cytoplasm and

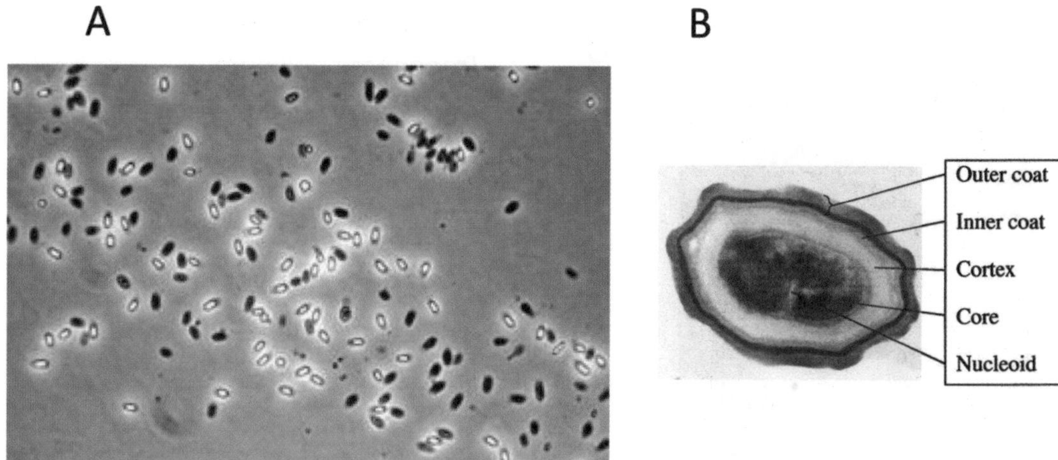

Figure 1-12 A. Spores of *Bacillus cereus*. The spores are highly refractile compared to the vegetative cells. Credit: Thomas Rosnes, Nofima (Norway). **B.** Cross-section of a *Bacillus subtilis* spore. Note the multiple protective layers. Credit: S. Pankratz, from (Nicholson, 2000).

highly condensed DNA, surrounded by multiple protective layers (Fig. 1-12B). Spores can survive for long periods. In fact, it has been claimed that viable spores were recovered from a bee encased in amber 25 million to 40 million years ago (Cano and Borucki, 1995). Streptomycetes also produce spores (arrow, Fig. 1-11D). These are called exospores because they develop as a long chain at the end of specialized cells rather than within the cell.

The cell envelope

KEY POINTS

- Most of the bacteria we encounter belong to two phyla, the *Firmicutes* and *Proteobacteria*.
- The *Firmicutes* have a thick cell wall made of peptidoglycan; the cell wall of the *Proteobacteria* is also peptidoglycan but much thinner.
- The cell wall prevents the cell from bursting when there is an influx of liquid by osmosis.
- The *Proteobacteria* have a complex structure, the outer membrane, surrounding the cell wall.
- Lipid A in the outer membrane is a potent inducer of inflammation.
- With the new technique of cryo-electron microscopy, the different parts of the cell envelope can be visualized.

Although there are many phyla of bacteria, most of the familiar bacteria we encounter belong to two of these, the ***Firmicutes*** and the ***Proteobacteria***. The bacilli, clostridia, lactobacilli, staphylococci, and streptococci are all members of the *Firmicutes*. The *Proteobacteria* is a very large phylum and includes most of the species that have been characterized to date. Falling within this group are *E. coli*, *Salmonella*,

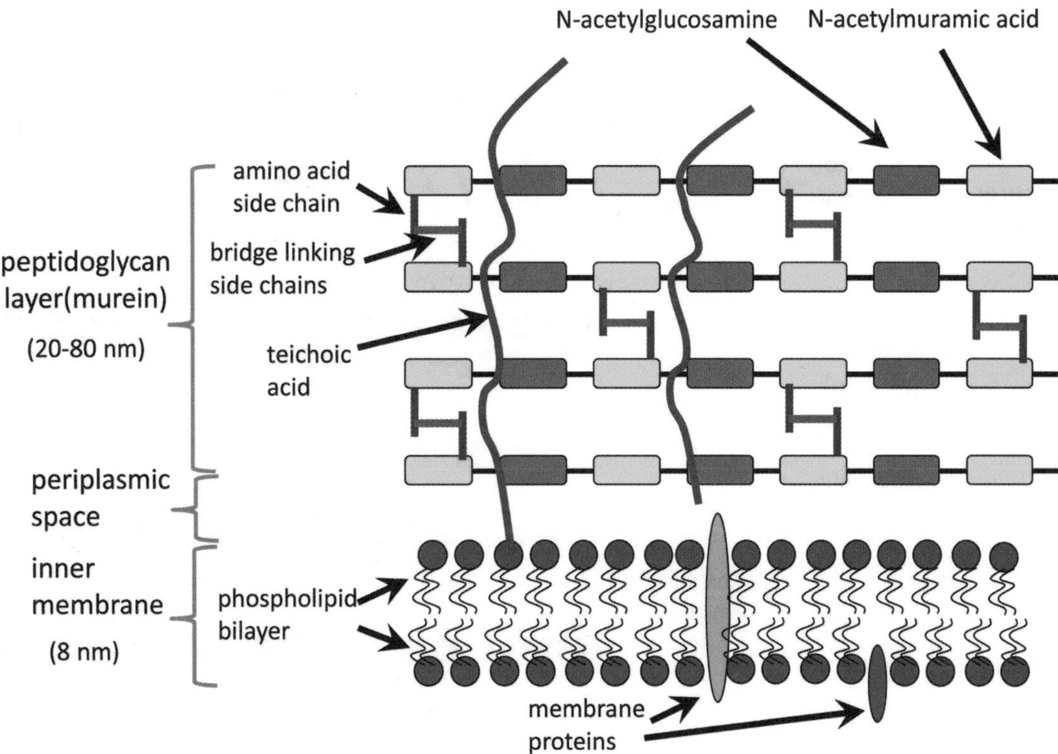

Figure 1-13 Cell envelope of the *Firmicutes.*

Pseudomonas, as well as many others. For reasons that will be clear later, the *Firmicutes* are often referred to as Gram-positive bacteria, and the *Proteobacteria* as Gram-negative.

There are many differences at the molecular level between the *Firmicutes* and *Proteobacteria*, indicating that these phyla diverged a very long time ago, near the beginning of bacterial evolution. One major difference between the two phyla lies in the structure of the **cell envelope**, the layers of molecular structures surrounding the cytoplasm. While both the *Firmicutes* and *Proteobacteria* have a symmetrical **cell membrane** composed of two oppositely facing but otherwise similar layers of phospholipids, the cell envelopes are otherwise very different. The *Firmicutes* have a thick cell wall, made up of **peptidoglycan**, alternating subunits of *N*-acetylglucosamine and *N*-acetylmuramic acid (Fig. 1-13). The *N*-acetylmuramic acid subunits have side chains, composed mostly of unusual amino acids, that cross-link adjacent strands of peptidoglycan, giving the cell wall rigidity in the second dimension. The result is a sturdy if porous mesh with a thickness up to 80 nm (1 nm = 10^{-9} meters). The cell wall forms a mesh around the cell that protects it from osmotic stress. Bacteria are sometimes in a hypotonic environment. This means that the concentration of all solutes outside the cell is less than inside the cell, resulting in the osmotic flow of water into the cell. The resulting turgor would rupture the membrane, but this force is counteracted by the rigidity of the cell wall, thus allowing

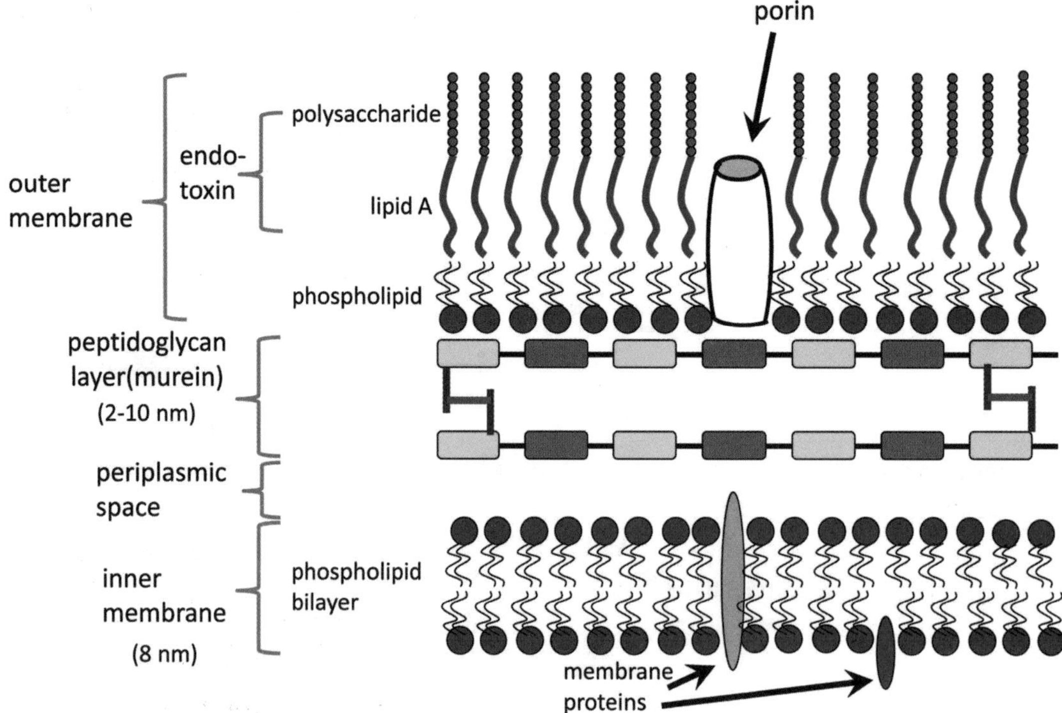

Figure 1-14 Cell envelope of the *Proteobacteria*.

the cell to retain its shape. This important function is revealed when cells are exposed to penicillin. In order to be effective, the cell wall must be linked in two dimensions. Penicillin prevents cross-linking of adjacent peptidoglycan strands. The result is an influx of water and lysis of the cell.

The structure surrounding the cell of a typical proteobacterium is more complicated. These cells also have a peptidoglycan cell wall, but it is very thin, about 2 nm (Fig. 1-14). The cell membrane and wall are separated by the **periplasmic space**. The periplasmic space is not empty: it is the home of many enzymes secreted through the cell membrane. Surrounding the cell wall of *Proteobacteria* is the outer membrane. The outer membrane is different from the symmetrical cell membrane (Fig. 1-14). The outer membrane is asymmetrical, with only the inner layer made up of phospholipid. The outer layer contains densely packed acidic lipid, attached to polysaccharide chains that extend outward into the environment. The major lipid component is highly conserved and is called **lipid A**. Lipid A and the polysaccharide chains on the cell surface are together called the **bacterial endotoxin**. Lipid A is a potent inducer of inflammation in humans. There are many proteins in the outer membrane; prominent among them are the **porins** (Fig. 1-14), barrel-shaped protein structures that allow the passage of hydrophilic molecules through the hydrophobic membrane.

The different layers of the cell envelope are too small to be visualized by light microscopy. Given this limitation, electron microscopy would seem like an attractive

firmicutes

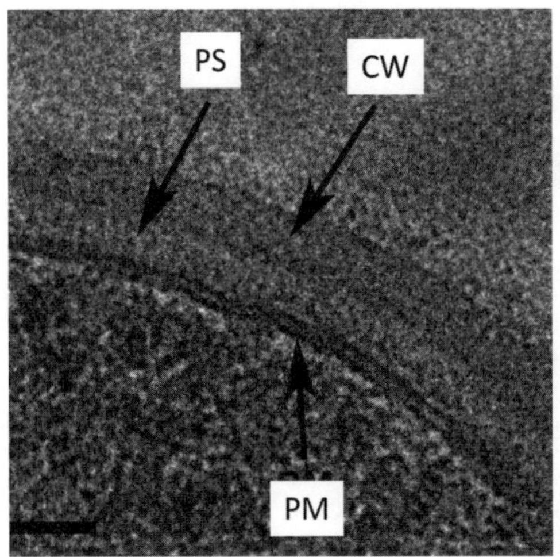

proteobacteria

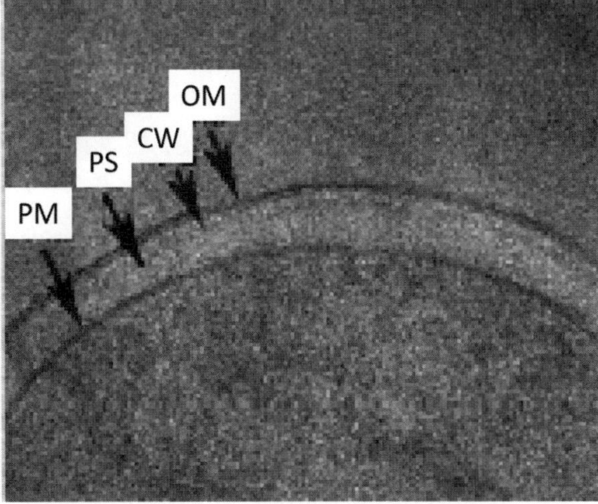

Figure 1-15 Cell envelopes visualized by cryo-electron microscopy. Modified from Beveridge, 2006. (Left) Firmicute *Staphylococcus aureus*. Credit: Valerio Matias. (Right) Proteobacterium *Pseudomonas aeruginosa*. PS = periplasmic space, PM = plasma membrane, CW = cell wall, OM = outer membrane.

alternative for researchers. In the electron microscope, an accelerating beam of electrons is used instead of light and focusing is achieved with magnetic instead of glass lenses. The wavelength of the electron beam is much smaller than the wavelength of light, which means that much higher resolutions are possible. In fact, the resolving power of the electron microscope is about 10^5 times greater than attainable with a light microscope and can resolve structures to nearly the atomic level. The problem is that the highly energetic electron beam required for large magnifications greatly damages biological specimens, obliterating or at least modifying the structures we wish to observe. However, this problem has now been largely overcome by the technique of cryo-electron microscopy. Here the specimen is frozen to very low temperatures, which reduces the amount of damage from the electrons while preserving the structural integrity of the specimen. The components of the cell envelope for *Firmicutes* and *Proteobacteria* have recently been nicely visualized by cryo-electron microscopy (Fig. 1-15).

Lab Two

1. *Streak the growth from a swab plate for isolated colonies.*
2. ***Select different isolated colonies as pure cultures.***
3. ***Characterize the microorganisms from each pure culture.***
4. *Identify the different bacteria at each site.*
5. *Determine which sites are contaminated with the microorganism causing disease.*
6. *Determine if these are the sites where disease is occurring.*
7. *Decide whether hospital items are the likely source of the disease-causing bacteria.*

Learning Outcomes

After this lab, students will be able to:

- **a.** Describe the morphology of bacterial colonies.
- **b.** Describe the morphology of a bacterial cell.
- **c.** Do a Gram stain.
- **d.** Identify optimum growth temperature based on colony size.

I. Describe the colony morphology of the unknown

Using the TSA plate containing fresh colonies of your unknown, describe the colony morphology, indicating those features that make the organism appear different from the other isolates in the sample. View your selected colony directly and with a stereomicroscope, if available. Colony morphology and color can be a good indicator of the identity of the bacteria: note, for example, the striking differences between colonies of *Bacillus cereus* and *E. coli* (Fig. 1-7).

Use the chart (Fig. 1-9) to describe three aspects of your colony: form, elevation, and margin. In addition, describe colony texture and color. Record your observations in your notebook.

SAFETY Never look directly at a UV light source. Always wear goggles designed to block UV.

Notice if there is a colored pigment around the colony. It is also worthwhile to look at your colonies under UV illumination. Remove the top of the petri dish and illuminate the surface of the medium with a long-wave, hand-held UV lamp. Do this in a darkened area (complete darkness is unnecessary) and put on UV goggles before turning on the lamp. Compare the appearance of your colonies with those isolated by others in the lab, and with a plate containing a fluorescent strain of *Pseudomonas* (provided). Several species of *Pseudomonas* produce pigments that diffuse from the cell, and these are often fluorescent (Fig. 1-16).

Figure 1-16 Streak plate of a *Pseudomonas* strain under visible and UV illumination.

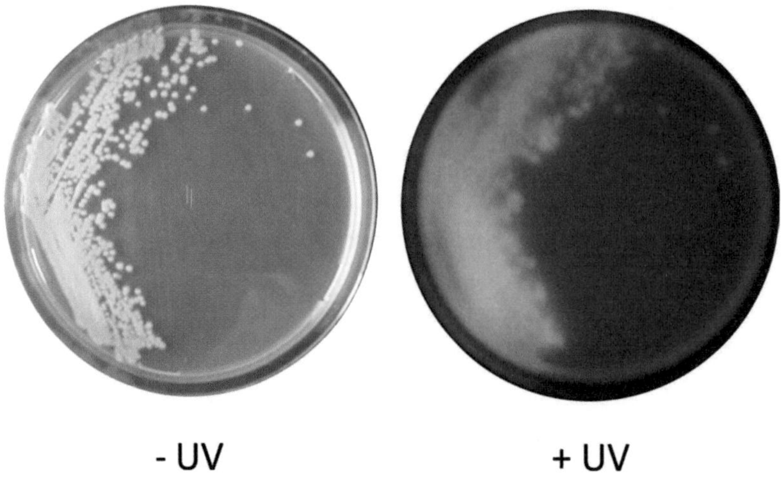

Serratia marcescens *Staphylococcus epidermidis*

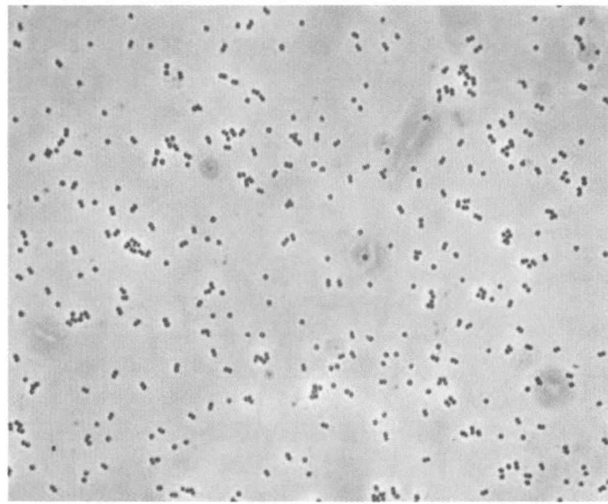

Figure 1-17 *S. marcescens* (left) and *S. epidermidis* (right) at 400 X magnification under phase contrast.

II. Describe the characteristics of an individual cell viewed under the microscope

Prepare a wet mount of cells from your isolated colony. Examine using the 40X and 1,000X objective lenses. If you have a phase-contrast microscope or if one is available in the lab, also view the cells with the phase-contrast filter in place. Determine whether the cells are cocci or rods and whether the cells are motile. For comparison, *S. epidermidis* and *B. cereus* cultures will be provided, and these should be examined as well. The cells of *S. epidermis* are round (coccoid) and nonmotile. *B. cereus* cells are rod-shaped and motile. *B. cereus* also produces spores, which can be easily distinguished using phase contrast. The spores are highly refractile and will appear bright (Fig. 1-12A). The spores develop inside regular cells, so look for their presence there as well.

Photographs of rod-shaped *S. marcescens* and coccoid *S. epidermidis*, viewed with a student microscope under phase contrast at 400X magnification, are shown in Fig. 1-17.

III. Determine the optimum temperature for growth

Starting from your isolated colony, restreak for isolated colonies on two TSA plates. Incubate one at 30°C and the other at 37°C. After 2 days, the plates will be removed and placed in the refrigerator (4°C).

IV. Determine if the unidentified microorganism is Gram positive or Gram negative

Since the *Firmicutes* and *Proteobacteria* form two major and distinct groups of bacteria, it would be very useful to have a rapid way of distinguishing them. In 1884, the Danish microbiologist Hans Gram published a procedure that would do

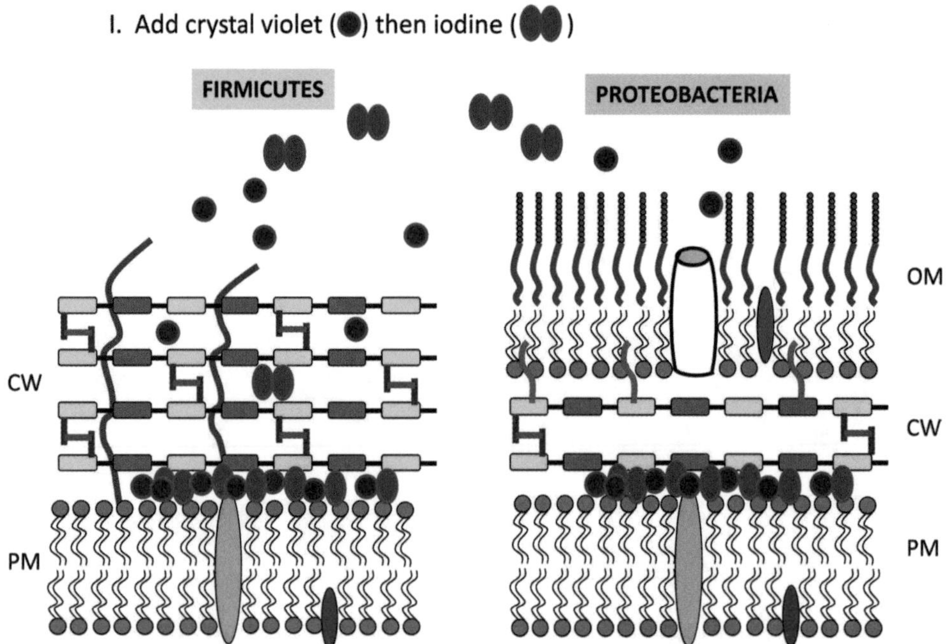

Figure 1-18 Gram stain, step 1. Crystal violet penetrates the cell wall. Subsequent addition of iodine solution results in dye complexes at the inner membrane of the cell.

just that. Although Gram was unsure whether his procedure would have lasting value, the Gram stain has been an important tool in microbiology for over 100 years.

The basis of the Gram stain lies in the difference in the thickness of the cell wall for *Firmicutes* and *Proteobacteria* (Fig. 1-13 and 1-14). There are minor variations, but the basic procedure has remained practically the same for years. The cells are first stained with crystal violet, a dye that easily passes through the peptidoglycan layer of both *Firmicutes* and *Proteobacteria* and mostly comes to occupy the space between the cell wall and cell membrane (Fig. 1-18). Importantly, the dye does not pass through the membrane into the cytoplasm of the cell. Gram's iodine solution is then added as a mordant, meaning it complexes with the dye and inhibits diffusion away from the cell. The next and critical step is treatment of the stained cells with ethanol. Ethanol quickly dehydrates the thick cell wall of *Firmicutes*, causing it to become less permeable and trapping the dye complex (Fig. 1-19). As a result, these cells remain stained and appear purple. In contrast, ethanol dissolves the outer membrane of *Proteobacteria*, diffuses through the thin peptidoglycan layer, and washes away the crystal violet-iodine complex (Fig. 1-19). Thus, the cells are decolorized. At this point, the proteobacterial cells will appear uncolored or faintly purple. However, the cells of *Firmicutes* will be deep purple due to retention of the dye complex. To improve the visibility of the *Proteobacteria* under the microscope, cells are then stained with a second dye (safranin is commonly used) (Fig. 1-20). The *Firmicutes* cells will remain purple, whereas the cells of *Proteobacteria* become pink and are thus more easily visualized (Fig. 1-21).

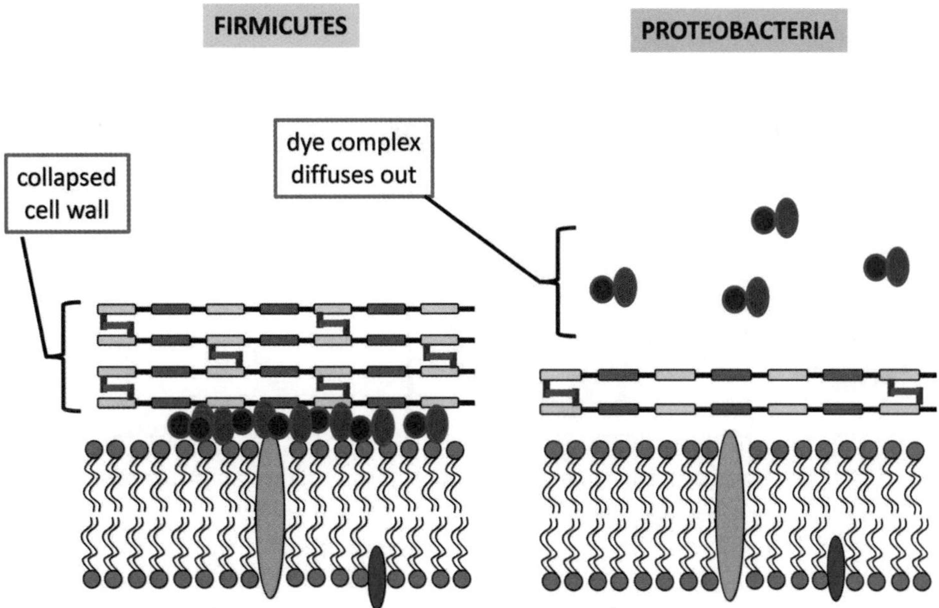

Figure 1-19 Gram stain, step 2. Ethanol dehydrates cell wall. Dye complex diffuses through thin cell wall of *Proteobacteria* but is retained by the thick collapsed cell wall of *Firmicutes.*

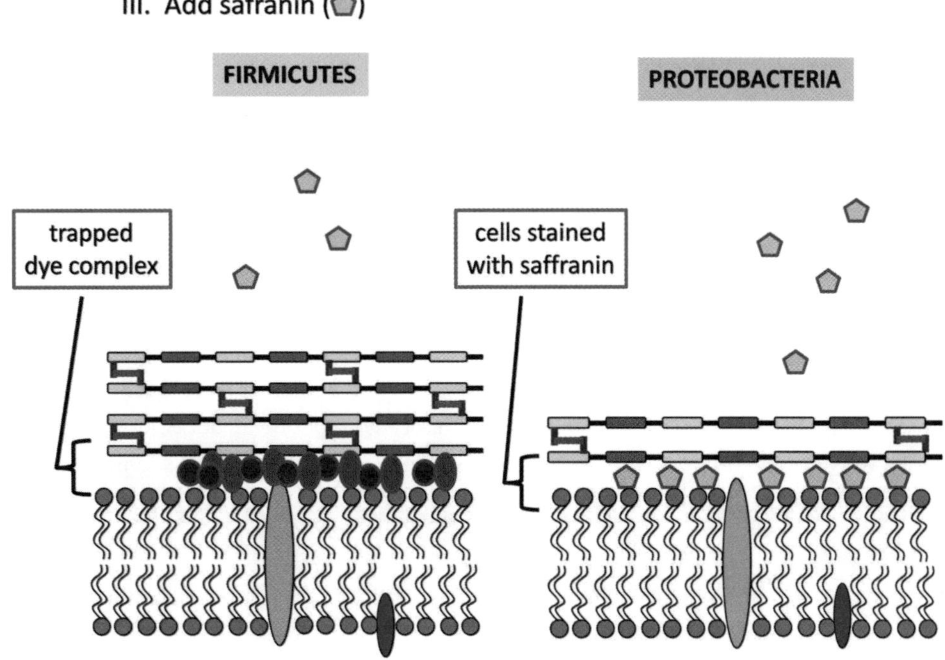

Figure 1-20 Gram stain, step 3. Safranin is added, which re-stains the now nearly colorless proteobacteria. Firmicutes cells remain stained with the crystal violet—iodine complex and safranin has no effect on their color.

Gram-positive

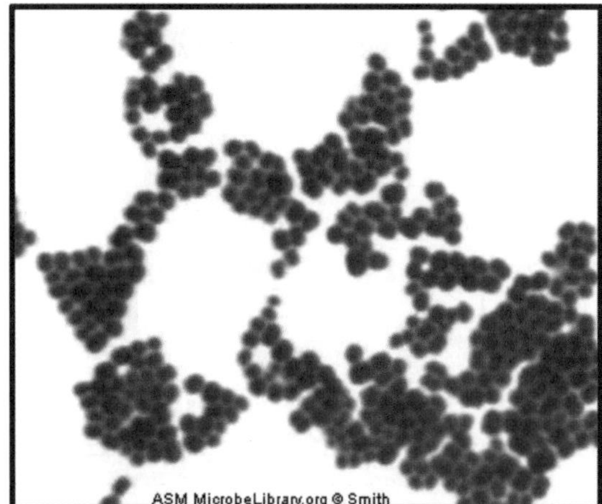

Gram-negative

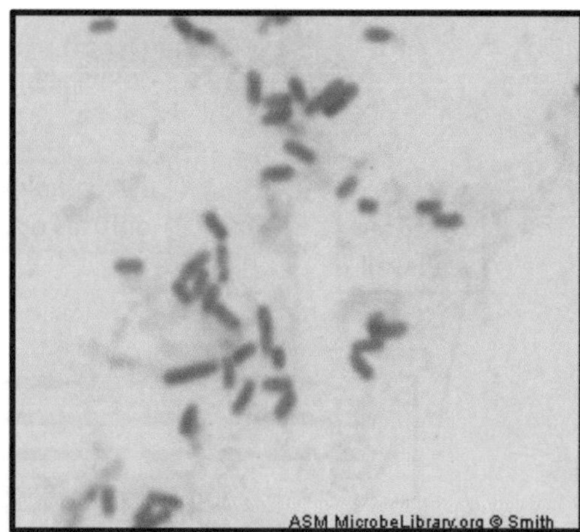

Figure 1-21 (left) Gram-positive *Staphylococcus epidermidis.* (right) Gram-negative *Escherichia coli.* ASM MicrobeLibrary. Credit: Ann C. Smith, University of Maryland.

PROCEDURE

Doing a Gram stain

Your instructor will demonstrate the technique for Gram-staining.

SAFETY Crystal violet will stain the skin and clothing. Always wear disposable gloves while doing a Gram stain and dispose of staining materials in the designated area. A lab coat or apron is advisable.

Stain your unknown and two controls, *S. epidermidis* (Gram positive) and *E. coli* (Gram negative), which will be provided.

1. Obtain two clean slides (no fingerprints). Divide each slide in half with a wax pencil (do not use a liquid marker) and then label: one half of each slide will contain your unknown and the other half one of the controls. On each half place a drop of water by using a sterile inoculating loop.
2. Transfer a small amount of cells from a fresh colony (1 to 3 days) to each of the droplets. A sterile wire needle is best for this—you do not want too many cells: looking at individual cells will give the most reliable result. Use the needle to distribute the cells uniformly by gentle mixing. Clumps of cells in the droplet will not stain properly. You should aim for droplets that become just slightly turbid.
3. Allow the liquid on the slides to air dry. Do not use a flame to hasten drying. Heating will cause the cells to burst.
4. For this and subsequent steps, hold the slide by using a clothespin, gripping the slide at one of the short ends. Pass each slide through the flame of your burner to fix the cells to the slide. The side containing your cells should be face up. Do this *quickly* three times. The areas of the slide containing bacteria should be warm but not hot.

SAFETY Remember, never leave an open flame unattended.

5. When the slide returns to room temperature, flood the slide with crystal violet stain (be sure you are staining the side that contains the bacteria). Do the staining over a tray designated for this

purpose: crystal violet stains most surfaces, including your clothing, lab benches, and sinks. Allow the stain to remain on the slide for about a minute; then, using a squeeze bottle, gently rinse the slide with water.

6. Add the iodine solution and let sit for 1 minute. Again wash with water. Place the slide on a paper towel (sample side up) to remove excess liquid.
7. Remove unbound crystal violet by rinsing the slide with 95% ethanol. Add the ethanol dropwise and continue until the ethanol draining from the slide into the tray is no longer colored. This should take about 10 seconds.
8. Wash the slide again with water and place on a paper towel.
9. Stain with safranin for 1 minute, then again rinse for a few seconds with water. Let dry on a paper towel as before.
10. Observe the cells under the microscope with the 100X oil immersion lens, if available, or with the 40X objective (no phase contrast). It is not necessary to use a coverslip. Record and interpret your observations.

Spores are not stained in the Gram procedure. The vegetative cells will stain normally, but the spores might appear Gram negative since they do not take up the stain (Fig. 1-22). Make sure you were not misled by the presence of spores in the sample.

Make sure that the controls stained properly before drawing conclusions about your unknown.

Some common problems:

1. No cells. Cells were probably not adequately fixed to the slide and were washed off during processing.
2. Cell fragments instead of whole cells. Slide was overheated during fixation.
3. Gram-negative control appears Gram positive. Insufficient destaining with ethanol.

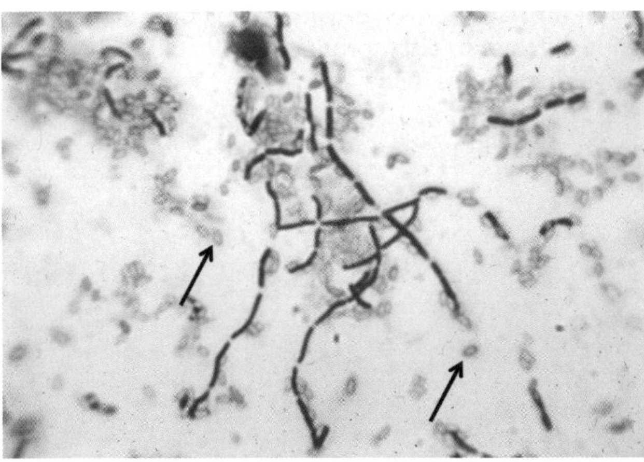

Figure 1-22 Gram-stained *Bacillus* species. Notice that the spores (arrows) are only lightly stained. Modified photograph from ASM MicrobeWorld (Wistreich collection).

4. Gram-positive control appears Gram negative. (i) Destaining time was too long or (ii) cells from a very old colony were used.
5. Cells appear both Gram positive and Gram negative. Too many cells added; cells in clumps not uniformly stained.

❖ Preparation for next lab

One to two days before your next lab session, retrieve your plates from the refrigerator. Looking at the sizes of the colonies on the two plates, decide whether the optimum growth temperature is 30 or 37°C and record the result. Restreak from the plate showing the best growth onto a new TSA plate and incubate at its optimum temperature for the next lab.

QUESTIONS

1. In the Background section for colony morphology, *S. marcescens* is used as an example. With this information, can you guess the growth temperature for the plate in Fig. 1-5? **(B1)**
2. Most common bacteria grow by cell division. Do you see any indication of this in the *S. marcescens* panel, Fig. 1-17? **(B1)**
3. While performing a Gram stain on your controls, you and your lab partner forget to add the counterstain safranin. What will be the color of *S. epidermidis* cells when viewed with the 100X oil immersion lens? *E. coli*? **(B2)**
4. You will need to prepare a fresh plate with isolated colonies for the next lab. Why should you use the colony you selected on the isolation plate and not the one on the original swab plate? **(B2)**
5. Old cells should not be used in the Gram procedure because Gram-positive cells can appear Gram negative. What does this tell you about the integrity of the cell wall when the cells are old? **(B5)**
6. The enzyme lysozyme, which can degrade peptidoglycan, is found in human tears. Why do you think tears are more effective as a defense against Gram-positive bacteria than against Gram-negative bacteria? **(B6)**
7. You are interested in studying soil microbes. You obtain a sample, streak for isolation, and incubate your plate for 6 days. The plate has a lot of growth; you decide to pick a large colony directly from your original plate for Gram staining. When you view your sample, you see both purple and pink cells, each slightly different in cell morphology. What likely happened? Design a simple experiment to test your hypothesis. **(B6)**

Lab Three

BACKGROUND

Modes of energy generation in bacteria

KEY POINTS

- Energy is generated by the flow of electrons from one compound to another one having a higher affinity for electrons.
- In substrate-level phosphorylation, the major energy carrier ATP is synthesized from ADP by the transfer of high-energy phosphate groups from different metabolic intermediates.
- Glycolysis is a central metabolic pathway where ATP is generated by substrate-level phosphorylation.
- ATP can also be synthesized by ATP synthase, where the source of energy is an electrically charged membrane.
- Respiration refers to the passage of electrons down the electron transport chain to generate the charged membrane required by ATP synthase.
- Many bacteria and almost all eukaryotic organisms use oxygen as the terminal electron acceptor during respiration because of its high affinity for electrons, but bacteria can use other terminal acceptors as well.
- Oxygen is highly reactive and can damage the cell by generating free radicals. Bacteria that use oxygen as a terminal acceptor have enzymes that destroy the reactive species.
- Some bacteria can generate all their energy by substrate-level phosphorylation. This type of energy generation is called fermentation.
- Glycolysis is often used to generate ATP during fermentation.

Energy for metabolism is generated by electrons passing from one chemical compound to another. This flow is due to differing affinity for electrons: compounds with low affinity will transfer electrons to compounds with high affinity. The donating compound is then said to be oxidized, and the receiving compound reduced. Of course, energy is not generated in a single step. Electrons are passed from compound to compound; when energy is released, it is converted to a form where it can do useful work. Where do these electrons ultimately come from? For microorganisms, there are three primary sources of electrons. One source is the oxidation of organic compounds. The second source of electrons is through photosynthesis. In this case, electrons in chlorophyll are excited by light and then returned to chlorophyll or passed to other compounds in the cell, in both cases with the release of energy. The third primary source of electrons is inorganic compounds. Unlike the other two sources, which are familiar to us because they are used by animals and plants, this source is unique to microorganisms. Bacteria in this group are less obvious, but they are extremely important. We mentioned previously that the conversion of NH_4^+ to nitrates by bacteria is essential for maintaining the fertility of the soil. This conversion is due to soil microorganisms that generate energy by oxidizing ammonium to nitrite and nitrite to nitrate, all inorganic compounds.

The energy of electron transfer is captured by cells in two ways. The first is by **substrate-level phosphorylation**: as compounds are oxidized, the resulting energy is retained in the oxidized compound as a high-energy phosphate bond. This energized phosphate is then transferred to adenosine diphosphate (ADP) to generate adenosine triphosphate (**ATP**). Once ATP is formed, the captured energy can drive many metabolic reactions in the cell. The second, and much more powerful, mechanism of generating ATP is through the creation of an unequal concentration of ions across the cell membrane. The charge imbalance essentially acts as a battery, providing the energy for ATP synthesis.

Bacteria encountered in the health sciences generate energy by the oxidation of organic compounds. An (admittedly oversimplified) overview is shown in Fig. 1-23. A wide variety of organic compounds are taken up, and many of these converted by different biochemical pathways into the central metabolite glucose. Glucose is then oxidized to generate energy for the cell. The glucose is first partially oxidized to pyruvate, usually by **glycolysis**, a central metabolic pathway, but some bacteria

Figure 1-23 The fate of electrons generated during the complete oxidation of glucose.

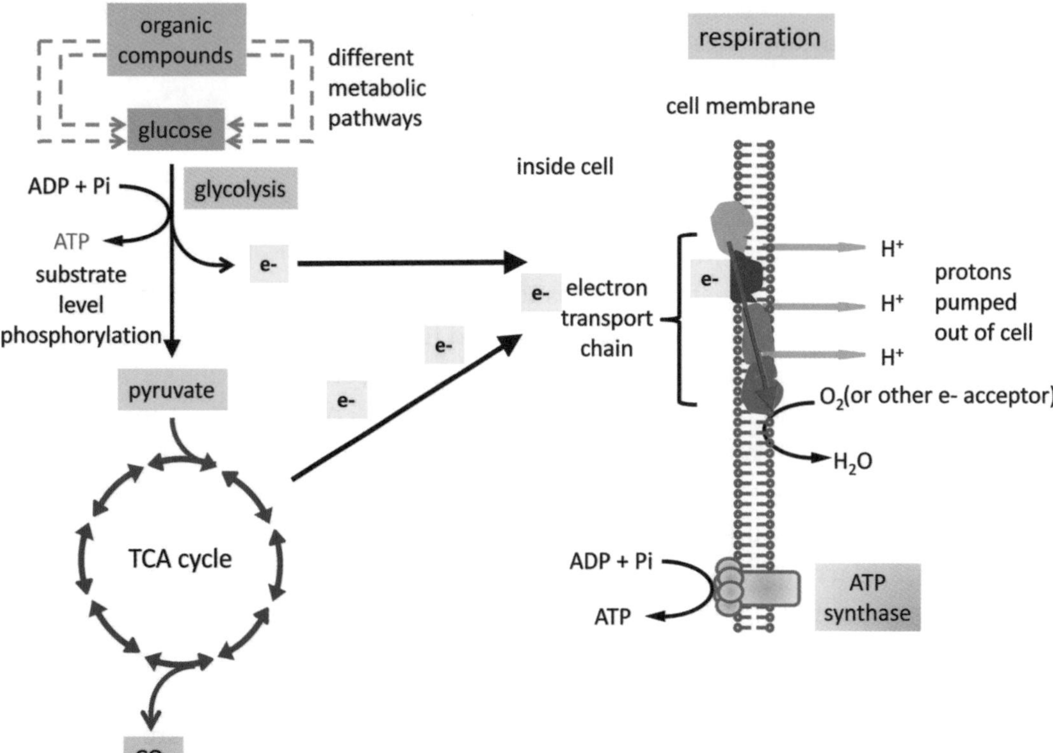

use other pathways. Pyruvate then enters the tricarboxylic acid (TCA) cycle, where it is completely oxidized. The electrons generated by oxidation are transported down a set of compounds that are mostly within the membrane. As the electrons pass down this **electron transport chain**, energy is released that is used to drive hydrogen ions out of the cell, so that there are more protons outside the membrane than in the cell. This difference in charge is the source of energy that drives **ATP synthase**, a molecular machine embedded in the membrane and capable of squeezing ADP and phosphate together to form ATP.

Although electrons passing down the electron transport chain lose energy, they still have to wind up somewhere. This "somewhere" is called the terminal electron acceptor. Almost all animals use oxygen as the terminal acceptor, and many bacteria can or must use oxygen as well. The generation of energy by the electron transport chain with oxygen as the final electron acceptor is referred to as **aerobic respiration**. Oxygen is a good terminal electron acceptor because it has a very high affinity for electrons, thus maximizing the yield of energy. In fact, about 1.5 billion to 2.5 billion years ago, precursors to eukaryotic cells adopted aerobic respiration by incorporating bacteria using this mode of energy generation. The remnants of these bacteria, still capable of respiration, are the mitochondria.

Some bacteria depend on **anaerobic respiration** for the generation of energy: that is, they use terminal electron acceptors other than oxygen. Denitrifying bacteria in oxygen-depleted soil and water can use nitrate (NO_3^-) as the terminal acceptor, converting it to nitrite (NO_2^-) and nitrogen gas (N_2). Other bacteria use different electron acceptors such as sulfur, which is reduced by the captured electrons to the foul-smelling H_2S. Anaerobic respiration is not a major mode of energy generation for bacteria that cause disease. Nevertheless, it is important to keep in mind the metabolic versatility of bacteria, including anaerobic respiration, in studying disease processes. For example, *E. coli* can generate energy from aerobic respiration but can also use nitrate as the terminal acceptor when oxygen is unavailable. Nitrate is produced as a by-product of the inflammatory response in the gut, which is a relatively anaerobic environment. By using nitrate in anaerobic respiration, *E. coli* might have a growth advantage compared to other organisms in the gut. This might account for the higher proportion of *E. coli* in the gut of individuals with a chronically inflamed bowel, caused by conditions such as Crohn's disease (Winter et al., 2013).

Some species of bacteria do not contain an electron transport chain for active respiration. Other species, such as *E. coli*, are still able to grow when a suitable terminal electron acceptor for respiration is absent. In these circumstances, ATP is synthesized solely by substrate-level phosphorylation, a mode of energy generation called **fermentation** (Fig. 1-24). Many different compounds can be fermented by bacteria, but the conversion of glucose to pyruvate by glycolysis is a major pathway. Glycolysis includes two energy-generating steps and, compared to other fermentation

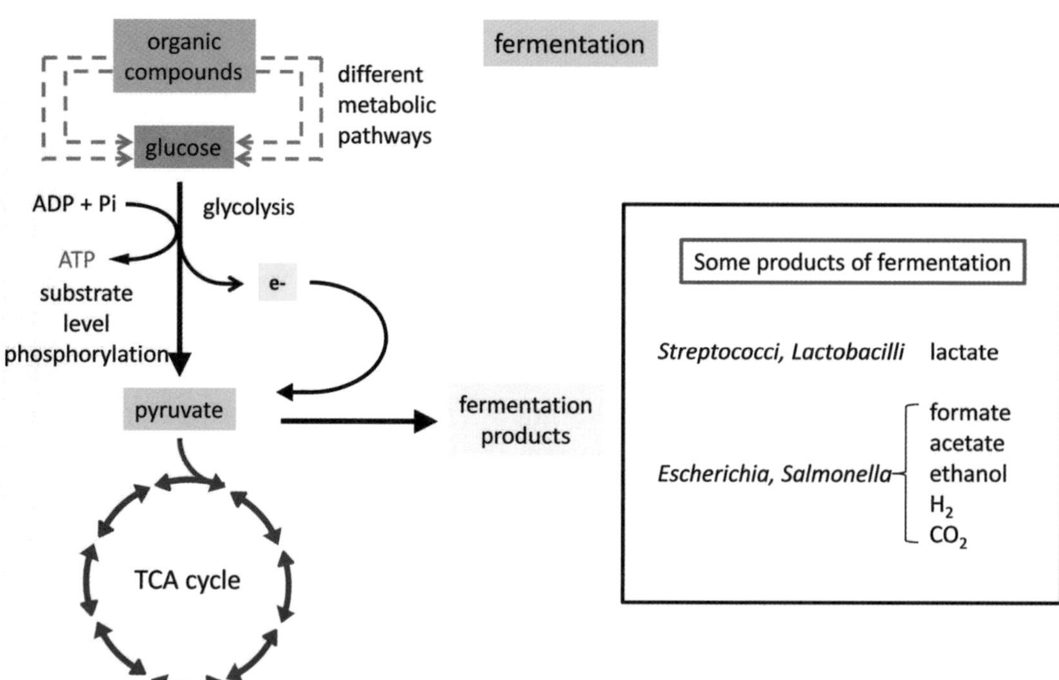

Figure 1-24 Energy generation by glycolysis during fermentation. Excess electrons are returned to pyruvate either to form lactic acid or to be further metabolized to a variety of different end-products. Part or all of the TCA cycle is present in fermenting microbes, but the cycle is run "backwards" for biosynthesis of key metabolites, rather than for respiration.

pathways, provides a higher energy yield. Still, 1 mole glucose yields a maximum of only 2 moles ATP during glycolysis, while a maximum of 36 moles of ATP are generated during aerobic respiration. (The actual values are less because the reactions are not 100% efficient.) You might think that fermenting bacteria are energy-starved and grow at a very slow rate as a result. *E. coli* and many other bacteria do grow more slowly when they are fermenting, compared to when aerobic respiration is active. The difference in rates is not as great as you might expect, and in fact, some fermenting bacteria grow at a rate comparable to respiring species. The reason is that although the yield of ATP is low, glycolysis is very rapid compared to respiration, with a large amount of glucose being consumed in a short time, so that the rate of ATP generation is high.

During fermentation, energy is still generated by the flow of electrons and these electrons still need to be finally deposited somewhere. While some are used during biosynthesis, most are simply discarded: in the case of glycolysis, by reducing pyruvate to one or more other compounds. *Escherichia*, *Salmonella*, *Klebsiella*, *Shigella*, and other members of the family *Enterobacteriaceae* (often called the "enterics") reduce pyruvate in a series of reactions to generate predominantly acetate, formate, ethanol, and the gases H_2 and CO_2. In contrast, the lactobacilli and streptococci, essential in cheese-making, produce only lactic acid by the direct reduction of pyruvate. Fermented foods are popular worldwide and indeed are often emblematic of a particular culture. This popularity is due to the complex flavor profile imparted by the products of fermentation. Cheese and other

superoxide dismutase: $2O_2^{-\cdot} + 2H^+ \rightarrow H_2O_2 + O_2$

superoxide radical hydrogen peroxide

catalase: $2H_2O_2 \rightarrow 2H_2O + 2O_2$

Figure 1-25 Reactions catalyzed by the enzymes superoxide dismutase and catalase.

fermented milk products are familiar examples; coffee beans, cocoa, sauerkraut, tofu, kombucha (fermented tea), and some pickled foods are also fermented. The Inuit peoples of the far north eat kiviak, young auks (a kind of seabird) fermented within the body of a seal. Not all bacteria are able to generate energy by fermentation of glucose. The majority of these rely on aerobic respiration. An important group in this category are the pseudomonads, although the important pathogen *P. aeruginosa* can survive in anaerobic environments by fermentation (Schreiber et al., 2006).

As stated above, oxygen has a strong affinity for electrons and is therefore a good terminal acceptor of electrons in respiration. However, there is a dark side to this: oxygen can capture extra electrons to form free radicals that can damage DNA and protein. Any bacterial species that can use oxygen for energy generation must cope with the potentially toxic effects of oxygen. This includes **obligate aerobes**, which must use oxygen for respiration (and cannot ferment), as well as **facultative anaerobes**, organisms that can respire aerobically but have at least one other mechanism for energy generation as well. *E. coli* is in the second group: it can respire both aerobically and anaerobically as well as acquire energy by fermentation. Aerobes and facultative anaerobes protect themselves from highly reactive forms of oxygen with the enzymes superoxide dismutase and catalase (Fig. 1-25), which rapidly break down the reactive species to water and molecular oxygen.

What about bacteria that do not use oxygen during energy production? **Strict anaerobes**, which grow only in the absence of oxygen, typically lack superoxide dismutase and catalase. For these organisms, oxygen is toxic. However, some bacteria generate energy anaerobically but are **aerotolerant**. These bacteria contain superoxide dismutase but not catalase. This makes sense: the superoxide radical is far more reactive than hydrogen peroxide.

Oxygen requirements, fermentation of sugars such as lactose and glucose, and the production of catalase are characteristics that are easily tested in the lab. Each test on its own is not enough for identification, but the results of all of these tests can generate a metabolic profile that, when used in conjunction with colony and cell morphology, can differentiate between genera and sometimes species.

Lab Three

1. *Streak the growth from a swab plate for isolated colonies.*
2. *Select different isolated colonies as pure cultures.*
3. ***Characterize the microorganisms from each pure culture (continuing).***
4. *Identify the different bacteria at each site.*
5. *Determine which sites are contaminated with the microorganism causing disease.*
6. *Determine if these are the sites where disease is occurring.*
7. *Decide whether hospital items are the likely source of the disease-causing bacteria.*

Learning outcomes

After this lab, students will be able to:

a. Transfer bacteria from solid medium to liquid broth.

b. Carry out and interpret the oxidase, catalase, glucose, and lactose fermentation tests.

I. Can the unidentified microorganism grow in the presence of bile salts and ferment lactose?

MacConkey medium is very useful because it is both *selective* (allows the growth of some microorganisms but not others) and *differential* (allows microorganisms to be distinguished based on their growth characteristics). This medium contains partially digested proteins as well as bile salts and neutral red, an indicator dye that turns red under acidic conditions. The Gram-positive *Firmicutes* are sensitive to bile salts and do not grow on MacConkey medium, whereas Gram-negative *Proteobacteria* are resistant to bile salts and generally grow well. Thus, the medium is a good way of confirming the results of the Gram stain. MacConkey medium is also widely used to test for fermentation of a particular sugar substrate, such as the milk sugar lactose. If the Gram-negative cells can ferment lactose, organic acids are produced that precipitate the bile salts and turn the medium a dark red due to the indicator dye (Fig. 1-26, right). If the *Proteobacteria* cannot metabolize lactose, then they utilize the broken-down proteins in the medium as an energy source. In this case, NH_3 is produced, the pH becomes more basic, and the cells and surrounding medium turn yellow or light pink, while the bile salts remain in solution (Fig. 1-26, left). *E. coli* and *Klebsiella* ferment lactose, but other common members of the *Enterobacteriaceae* and the pseudomonads do not.

Bacteria can also be tested for the fermentation of other sugars by using MacConkey medium. For example, sorbitol, another sugar, is also generally fermented by strains of *E. coli* but not by the highly pathogenic strain O157:H7. Thus, MacConkey medium, with sorbitol instead of lactose, can be used as an indicator medium during outbreaks of disease due to this dangerous strain of *E. coli*.

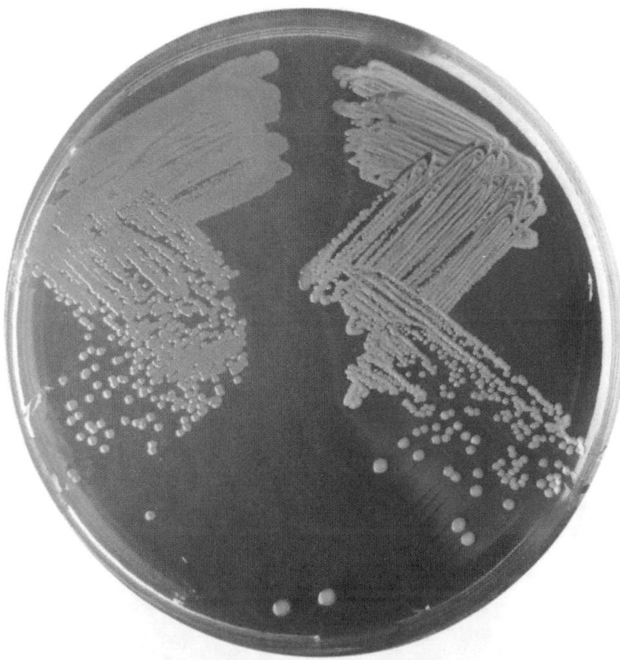

Figure 1-26 Growth on a MacConkey plate. (left) *Pseudomonas aeruginosa* negative for fermentation of lactose. (right) *E. coli*, a lactose-fermenter: note red-purple color of the medium and the precipitation of bile salts around colonies.

PROCEDURE

Streaking cells on MacConkey-lactose plates

As a group, streak the controls *Pseudomonas putida, E. coli*, and *S. epidermidis* on three MacConkey-lactose (MAC) plates. Working individually, streak your unknown on a MAC plate as well. Incubate the plate containing *Pseudomonas* at 30°C, the other controls at 37°C, and your unknown at its optimum growth temperature. Examine the plates after incubation for 24 to 48 hours and record the results in your notebook.

II. Can the unidentified microorganism ferment glucose?

Your unknown was able to grow in the presence of oxygen, so it cannot be an obligate anaerobe, which is poisoned by oxygen. However, it might be able to grow under anaerobic conditions (i.e., it might be a facultative anaerobe) by fermenting glucose. A culture tube containing broth medium, glucose, and phenol red is inoculated with bacteria and allowed to incubate. If the bacteria are able to grow, the medium will become visibly cloudy (turbid). Because the tube is not shaken to introduce air, growth is under semianaerobic conditions. If your unknown is able to metabolize the glucose in the tube by fermentation, acids are produced and the phenol red indicator dye turns yellow (Fig. 1-27). If your unknown does not ferment glucose but is still able to grow under semianaerobic conditions, it will metabolize the peptides in the complex medium. In that case, the medium will remain red, because the products of this metabolism are basic. However, the medium will still become turbid because of the bacterial growth (Fig. 1-27). Finally, if there is no growth, then the tube will remain clear and nothing can be concluded about fermentation, since the medium might not support growth of the microorganism under any circumstances.

Growth:	yes	yes	no
Glucose fermentation:	yes	no	--

Figure 1-27 The glucose fermentation test Credit: Seth James.

PROCEDURE

Glucose fermentation test

Transferring cells to liquid medium is not difficult, but any contamination would probably remain undetected since the liquid cultures of different bacteria often look the same. For this reason, good sterile technique is important.

1. Label one of the glucose fermentation tubes with your name, the date, and the microorganism. Do not label the cap.
2. Place the plate containing your unknown upside down on the bench, so that the cover is on the benchtop.
3. Sterilize your loop as before and let it cool. With your free hand, lift the bottom of the plate from the cover and touch a fresh colony of your unknown with the loop, then immediately return the plate to its cover.
4. Remove the cap from one of the tubes containing glucose fermentation broth. To do this, hold the tube firmly with your free hand and remove the cap using the fourth and fifth fingers of the hand holding the loop. Continue to hold the cap: do not place it on the bench. See **Technique Box 2** for the appropriate way to hold the cap.
5. Touch the loop to the liquid. You can move the loop around in the liquid for a few seconds to help dislodge the cells, but avoid touching the side of the tube with the handle. Replace the cap immediately.
6. Remember to sterilize the loop before placing on the bench.

TECHNIQUE BOX 2

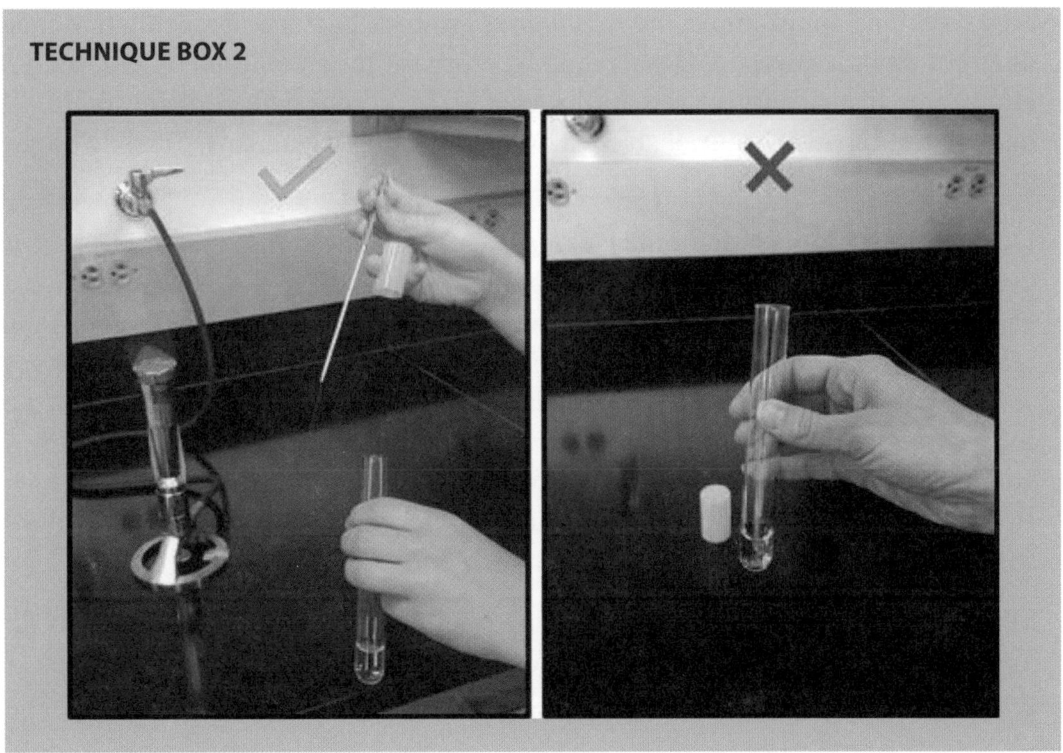

As a group, also prepare control tubes containing *P. putida* and *E. coli* from the plate cultures provided. Incubate *P. putida* at 30°C and *E. coli* at 37°C, the optimum growth temperature for each. Incubate your unknown at its optimum temperature as well. During this time do not shake or disturb the cultures, which would introduce air into the medium. After 48 hours, compare your result with the controls. To detect turbidity, it might be necessary to *gently* swirl the culture once to bring the cells up from the bottom of the tube. Glucose fermenters generally use glucose first but, when this carbon source is depleted, will resort to metabolizing peptides. As a result, the yellow medium will turn red over time as it becomes more basic, so do not wait longer than 48 hours. If returning to the lab at this time is not feasible, the instructor will photograph the results and post them on the course website. Remember, the test depends on good bacterial growth; if your cells do not grow well in the medium (and some bacteria do not), then nothing can be concluded, so look for turbidity.

III. Does the unidentified microorganism use cytochrome *c* during respiration (Gram-negative bacteria)?

The oxidase test indicates whether bacteria contain cytochrome *c* and the enzyme cytochrome *c* oxidase. All obligate aerobes use this complex in the electron transport chain to transfer electrons to oxygen (Fig. 1-23). Facultative anaerobes, including *E. coli*, generally use different cytochromes and corresponding oxidases to carry out this reaction. These bacteria are oxidase negative. Be aware, though, that there are exceptions: for example, the *Aeromonas* group of bacteria are facultative anaerobes but oxidase positive. However, for the organisms encountered here, the generalization that a positive test indicates a strict aerobe will probably hold. The oxidase test can also be unreliable with Gram-positive organisms, possibly because of the thick cell wall.

During respiration, cytochrome *c* oxidase catalyzes the transfer of electrons from cytochrome *c* to the electron acceptor oxygen (Fig. 1-28). In the oxidase test, a drop of the reagent tetramethyl-*p*-phenylenediamine (TMPD) is applied to cells. This reagent is colorless, but the oxidized derivative is blue. Cytochrome *c* has a greater affinity for electrons than TMPD, and as a result, electrons flow from TMPD to cytochrome *c*: the reagent turns blue because it is being oxidized. The reaction is a good example of how catalysis can occur in both directions. Although the flow of electrons is reversed compared to respiration, the enzyme cytochrome *c* oxidase is still required for the reaction to occur.

The cytochromes of most facultative anaerobes have less affinity for electrons than TMPD. As a result, electrons do not flow from TMPD to these cytochromes and the reagent remains colorless.

Figure 1-28 Electron flow during the oxidase test.

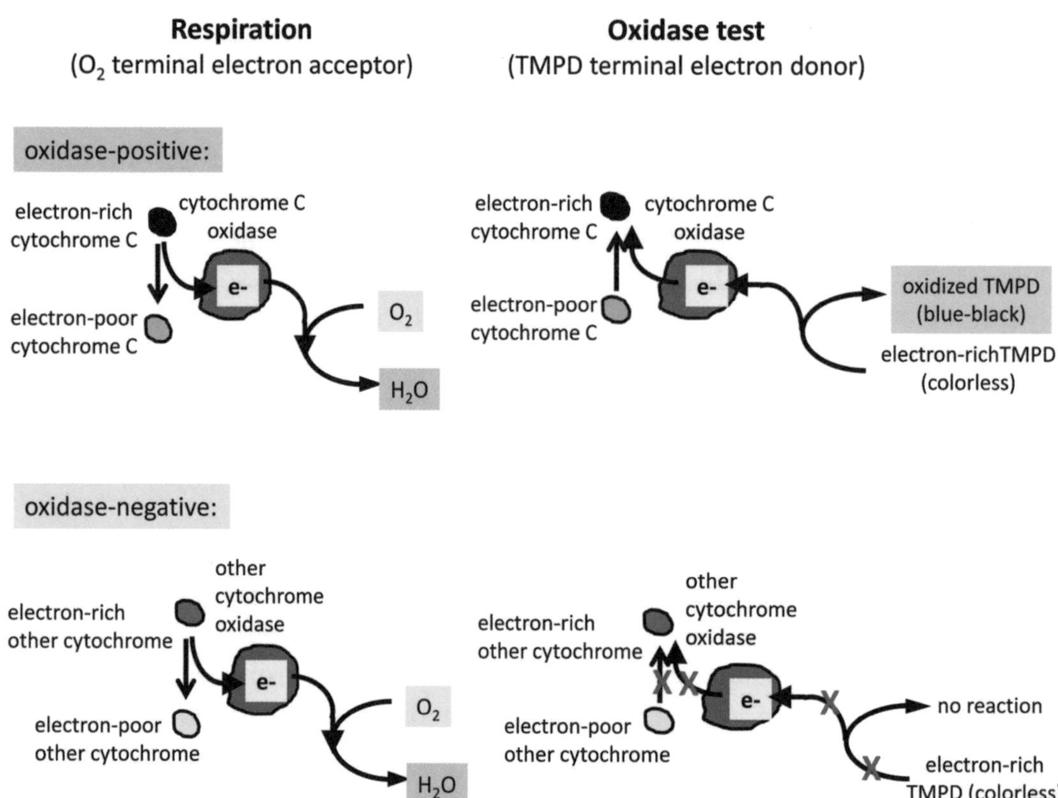

PROCEDURE

Oxidase test

Using a sterile toothpick, transfer cells from one or two fresh colonies onto a sterile filter paper strip. Spread the cells so that there is a uniform, circular patch on the paper. Place one drop of the reagent onto the cells and observe if there is a color change. In positive cases, the change should occur after 5 to 20 seconds (Fig. 1-29). Ignore any color change after a longer interval, which can be due to spontaneous oxidation of the reagent. For comparison, there will be plates containing freshly streaked *P. putida* (oxidase positive) and *E. coli* (oxidase negative) in the lab.

Figure 1-29 Oxidase test: (left) negative (right) positive. Modified from image at Slideshare.net.https://www.slideshare.net/drmalathi13/oxidase-test. Credit: Malathi Murugesan Wellcome Research Laboratory, Christian Medical College, Vellore India.

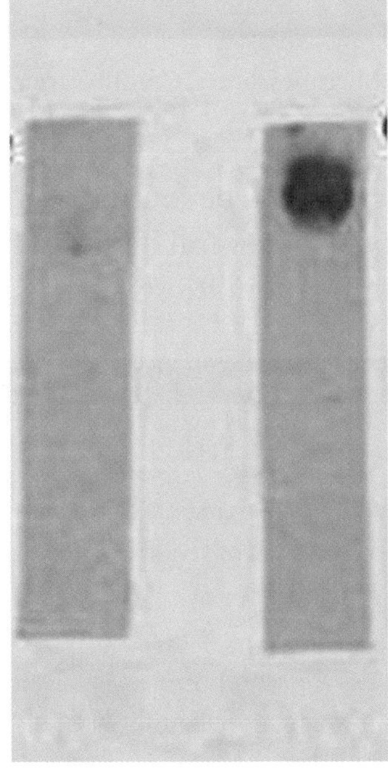

IV. Does the microorganism make catalase (Gram-positive bacteria)?

If the cell produces catalase, then H_2O_2 is converted to O_2 and H_2O (Fig. 1-30). Normally the amount of hydrogen peroxide is low and the oxygen produced is not readily detectable. In this test, a much greater amount of hydrogen peroxide is added to the cells and the production of O_2 is visualized by the appearance of bubbles (Fig. 1-30). The test is very useful for distinguishing between staphylococci and streptococci, two groups of Gram-positive cocci. Both groups of bacteria are able to survive in the presence of oxygen. Staphylococci are able to grow by aerobic respiration and are therefore catalase positive. Streptococci generate energy by fermentation but are aerotolerant. They contain superoxide dismutase but lack catalase. They are related to the lactobacilli (which are rod-shaped), and as mentioned

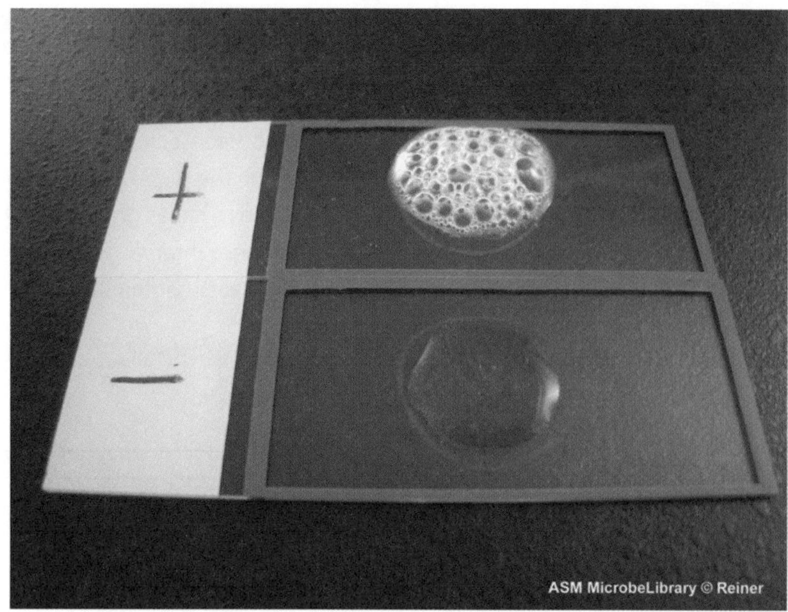

Figure 1-30 Catalase test. ASM MicrobeLibrary. Credit: Karen Reiner, Andrews University.

earlier, members of both groups can be found in fermented milk products. Species of streptococci are responsible for several diseases, from superficial infections such as "strep throat" to more serious conditions such as toxic shock syndrome.

PROCEDURE

Catalase test

Select a clean slide and divide into halves with a wax marker. Using a sterile toothpick (not your loop), transfer cells from a colony of your unknown onto one half of the slide and from a colony of the positive control (*S. epidermidis*) on the other half. Use a visible number of cells and apply to the slide with a circular motion so that the diameter of the patch is about 5 mm. Place a few drops of 3% H_2O_2 from a dropper bottle onto each patch. The positive control should start forming bubbles almost immediately. Examine whether your unknown is also releasing oxygen. No bubbles, or a small number of bubbles appearing after 20 seconds, is considered a negative result.

V. Is the microorganism motile?

You have probably drawn an initial conclusion concerning the motility of your organism by observing living cells with the microscope. Another way of testing for motility is to ask whether the bacteria can move through medium containing a low concentration of agar. One common method is to use a deep plug of this medium (Fig. 1-31, left). Cells are transferred into the plug with a straight wire. If the cells are motile, growth will extend into the agar away from the initial stab line. Non-motile bacteria will remain near the site of the inoculation and grow only along the stab line. The extent of growth is made easier to observe by adding a small amount of 2,3,5-triphenyltetrazolium to the medium. Initially colorless, this compound is reduced to a red product (formazan) by growing bacteria. The problem with this widely used approach is that most of the cells in the stab are growing semianaerobically, and this can reduce or inhibit motility. Instead, you will observe if the cell population swims in soft agar medium in a petri dish.

PROCEDURE

Soft agar motility assay

Using a sterile wire (not a loop), inoculate the center of the dish with your unknown and transfer to the optimum growing temperature. The agar medium is very soft: be careful moving the plate and incubate *cover side up*. As a group, inoculate two other motility plates with *E. coli* (motile) and *S. epidermidis* (nonmotile). Incubate these plates at 37°C. Examine the plates after 48 hours. If the cells are motile, they will swim outward from the stab and a circular "cloud" of bacteria will be seen (Fig. 1-31, right). Nonmotile cells will grow but will remain at the site of inoculation. Swimming outward from the inoculation site is an example of positive chemotaxis. As the bacteria deplete the nutrients locally, they move outward in search of a fresh food supply. Chemotaxis is investigated more fully in another challenge.

Figure 1-31 (left) Motility test, agar plug stabs. ASM MicrobeLibrary. Credit: Patricia Shields, Laura Cathcart. (right) Motility test in 0.3% agar.

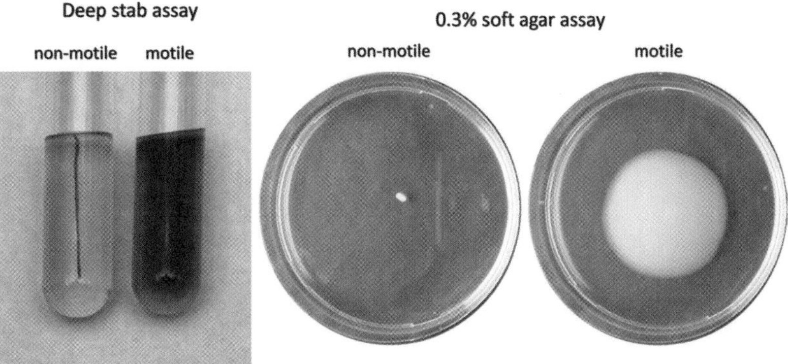

QUESTIONS

1. Correctly place the compounds "TMPD," "other cytochrome-oxidase complex," "O_2," and "cytochrome *c*-cytochrome *c* oxidase" according to their affinity for electrons at the positions A, B, C, and D in Fig. Q1-3. (B1)

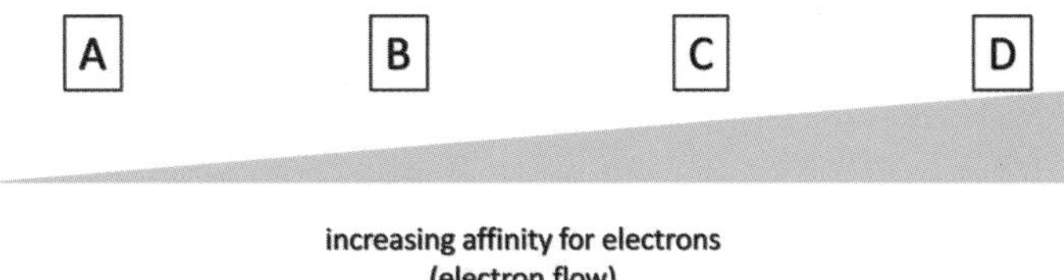

Figure Q1-3 Flow of electrons in oxidase test.

2. Suppose one of the organisms from the hospital samples grew well at 30°C but extremely poorly at 37°C. Explain why this organism is not a likely candidate for the one causing the hospital infections. **(B2)**
3. Explain why the aerotolerance of streptococci is an important factor in their ability to cause a wide variety of different diseases. **(B2)**
4. In the glucose fermentation test, sometimes the medium turns yellow but there is a pronounced red band at the interface between the medium and the air. State a hypothesis to explain this observation. How would you test your hypothesis? **(B6)**

Solving Challenge One

1. *Streak the growth from a swab plate for isolated colonies.*
2. *Select different isolated colonies as pure cultures.*
3. *Characterize the microorganisms from each pure culture.*
4. ***Identify the different bacteria at each site.***
5. ***Determine which sites are contaminated with the microorganism causing disease.***
6. ***Determine if these are the sites where disease is occurring.***
7. ***Decide whether hospital items are the likely source of the disease-causing bacteria.***

You are now ready to determine if the source of the microorganism causing disease is likely contaminated hospital items.

Working individually, fill out Table 1-1 for your unknown. In the "Comments" column indicate any tests that were ambiguous or inconclusive. Compare your results with Table 1-2, which gives the properties of several different genera that can cause hospital infections resulting from contamination.

After making the comparison, decide on the most probable genus of your unknown. Add your name and the proposed genus to Table 1-3, which will be posted online or in the lab. If the characteristics of your unknown do not match perfectly with the characteristics of any of these microorganisms, note the discrepancy.

Your instructor will now tell you the identity of the microorganism causing the illness and which samples are from infected areas and which are not. Decide whether the source of this microorganism is likely to be from contaminated hospital items.

❖ Preparing for Challenge Two

Meet as a group and look over the results for your sample (Tables 1-1 and 1-3). Select as candidates for sequencing two of the strains in your sample. If identification of an isolate is uncertain, or if you think that one of the isolates in your sample is causing the illness, then these would be good candidates. One to two days before beginning Challenge Two, purify each isolate on a separate TSA plate and incubate at its optimum temperature so that you will have fresh colonies for the challenge.

Table 1-1 The properties of your unidentified microorganism

Hospital sample number:_____

Test:	Result:	Comments:
Colony appearance		
Cell shape		
Spores		
Motility		
Optimum temperature		
MAC medium		
Glucose fermentation		
Oxidase		
Catalase		

Table 1-2 Properties of some common bacterial contaminants in hospitals

Genus	Representative species	Cell shape	Motility	Spores	Gram stain	Growth temp. °C	MAC	Glucose fermentation	Oxidase test	Catalase test
Pseudomonas	fluorescens	rod	yes	no	–	30	growth Lac–	no	pos.	ND
Acinetobacter	baumannii	short rod	no	no	–	25–30	growth Lac–	no	neg.	ND
Escherichia	coli	rod	yes	no	–	37	growth Lac+	yes	neg.	ND
Klebsiella	pneumoniae	rod	no	no	–	37	growth Lac+	yes	neg.	ND
Bacillus	subtilis	rod	yes	yes	+	37	no growth	yes	ND	pos.
Staphylococcus	aureus	coccus (cluster)	no	no	+	37	no growth	yes	ND	pos.
Streptococcus	pyogenes	coccus (chain)	no	no	+	37	no growth	yes	ND	neg.

ND = no data (test not done)

Table 1-3 Bacteria identified at each site in the hospital

Hospital sample	Class member	Proposed genus
1.		
2.		
3.		
4.		
5.		
6.		

CHALLENGE ONE QUESTIONS

1. Two groups of four students were assigned the same hospital sample. The bacteria below were identified in the sample by each member of the group:

   Group A:
   (two students): *E. coli*
   (two students): *Pseudomonas*

   Group B:
   (two students): *E. coli*
   (one student): *Pseudomonas*
   (one student): *Klebsiella*

   Using the tests in **Table 1-2,** how would you go about resolving the difference between the two results? **(B6)**

2. Do any of the following results rule out the hypothesis that the disease was due to contamination of the tested hospital items?
   a. All the samples, from both infected and uninfected areas, tested positive for the microorganism causing disease.
   b. The disease-causing microorganism was only found on one sample. This sample was from an area of the hospital experiencing infections.
   c. One of the microorganisms was found only on samples from infected areas in the hospital, but it was not the one identified by the laboratory as causing disease. **(B6)**

BIBLIOGRAPHY

Beveridge TJ. 2006. Visualizing bacterial cell walls and biofilms. *Microbe* **1:**279–284. https://www.asm.org/ccLibraryFiles/FILENAME/000000002339/znw00606000279.pdf.

Cano RJ, Borucki MK. 1995. Revival and identification of bacterial spores in 25- to 40-million-year-old Dominican amber. *Science* **268:**1060–1064. http://science.sciencemag.org/content/268/5213/1060.abstract.

Cupcakes and Cute Things. August 2009. http://cupcakesandcutethings.blogspot.com/2009_08_01_archive.html.

D'Amico S, Collins T, Marx JC, Feller G, Gerday C. 2006. Psychrophilic microorganisms: challenges for life. *EMBO Rep* **7:**385–389. https://www.ncbi.nlm.nih.gov/pmc/articles/PMC1456908/?tool=pmcentrez.

Dykhuizen D. 2005. Species numbers in bacteria. *Proc Calif Acad Sci* **56**(6 Suppl 1)**:**62–71. https://www.ncbi.nlm.nih.gov/pmc/articles/PMC3160642/?tool=pmcentrez.

Griffiths BS, Ritz K, Glover LA. 1996. Broad-scale approaches to the determination of soil microbial community structure: application of the community DNA hybridization technique. *Microb Ecol* **31:** 269–280. http://link.springer.com/10.1007/BF00171571.

Guinet F, Carniel E, Leclercq A. 2011. Transfusion-transmitted *Yersinia enterocolitica* sepsis. *Clin Infect Dis* **53:**583–591. http://cid.oxfordjournals.org/content/53/6/583.full.

Kallmeyer J, Pockalny R, Adhikari RR, Smith DC, D'Hondt S. 2012. Global distribution of microbial abundance and biomass in subseafloor sediment. *Proc Natl Acad Sci U S A* **109:**16213–16216.

Leboffe MJ, Pierce BE. 2012. *Brief Microbiology Laboratory Theory & Application*, 2nd ed. Morton Publishing, Denver, CO.

Lei Y, Oshima T, Ogasawara N, Ishikawa S. 2013. Functional analysis of the protein Veg, which stimulates biofilm formation in *Bacillus subtilis*. *J Bacteriol* **195:**1697–1705. http://www.ncbi.nlm.nih.gov/pubmed/23378512.

Luef B, Frischkorn KR, Wrighton KC, Holman HY, Birarda G, Thomas BC, Singh A, Williams KH, Siegerist CE, Tringe SG, Downing KH, Comolli LR, Banfield JF. 2015. Diverse uncultivated ultra-small bacterial cells in groundwater. *Nat Comm* **6:**6372. http://www.nature.com/ncomms/2015/150227/ncomms7372/full/ncomms7372.html.

Mora C, Tittensor DP, Adl S, Simpson AG, Worm B. 2011. How many species are there on Earth and in the ocean? *PLoS Biol* **9:**e1001127. http://journals.plos.org/plosbiology/article?id=10.1371/journal.pbio.1001127.

Morris RM, Rappé MS, Connon SA, Vergin KL, Siebold WA, Carlson CA, Giovannoni SJ. 2002. SAR11 clade dominates ocean surface bacterioplankton communities. *Nature* **420:**806–810. http://dx.doi.org/10.1038/nature01240.

National Institutes of Health. 2012. NIH Human Microbiome Project defines normal bacterial makeup of the body. National Institutes of Health, Bethesda, MD. http://www.nih.gov/news-events/news-releases/nih-human-microbiome-project-defines-normal-bacterial-makeup-body.

Nicholson W, Munakata N, Horneck G, Melosh H,Setlow P. 2000. Resistance of Bacillus endospores to extreme terrestrial and extraterrestrial environments. *Microbiol Mol Biol Rev* **64:** 548–572. doi:10.1128/MMBR.64.3.548-572.2000

Schreiber K, Boes N, Eschbach M, Jaensch L, Wehland J, Bjarnsholt T, Givskov M, Hentzer M, Schobert M. 2006. Anaerobic survival of *Pseudomonas aeruginosa* by pyruvate fermentation requires an Usp-type stress protein. *J Bacteriol* **188:**659–668. http://www.ncbi.nlm.nih.gov/pubmed/16385055.

Whitman WB, Coleman DC, Wiebe WJ. 1998. Prokaryotes: the unseen majority. *Proc Natl Acad Sci U S A* **95:**6578–6583.

Whitman WB, Woyke T, Klenk HP, Zhou Y, Lilburn TG, Beck BJ, De Vos P, Vandamme P, Eisen JA, Garrity G, Hugenholtz P, Kyrpides NC. 2015. Genomic Encyclopedia of Bacterial and Archaeal Type Strains, Phase III: the genomes of soil and plant-associated and newly described type strains." *Stand Genomic Sci* **10:**26. http://www.pubmedcentral.nih.gov/articlerender.fcgi?artid=4511459&tool=pmcentrez&rendertype=abstract.

Winter SE, Winter MG, Xavier MN, Thiennimitr P, Poon V, Keestra AM, Laughlin RC, Gomez G, Wu J, Lawhon SD, Popova IE, Parikh SJ, Adams LG, Tsolis RM, Stewart VJ, Bäumler AJ. 2013. Host-derived nitrate boosts growth of *E. coli* in the inflamed gut. *Science* **339:**708–711. http://science.sciencemag.org/content/339/6120/708.abstract.

challenge Two

Confirming the identification of a microorganism by sequencing the 16S rRNA gene

In Challenge One, you carried out an analysis of bacteria isolated from different sites in a hospital. You believe that you have identified the microbes present in the different samples. You now plan to confirm your results by determining the DNA base sequence for part of the 16S rRNA gene for two of the isolates from your sample. The sequence data will be compared with the 16S rRNA gene sequences of other bacteria to find the microorganism(s) with the most closely related sequence.

QUESTIONS BEFORE YOU BEGIN THE CHALLENGE

1. What assumption is being used here to identify the isolates?
2. What is the expected outcome if the hypothesis in Challenge One is correct?

Strategy for Challenge Two

1. *Use whole-cell PCR to amplify the 16S rRNA gene.*
2. *Verify by agarose gel electrophoresis that the PCR products are present.*
3. *Submit the PCR products for automated DNA sequencing at a core facility.*
4. *Analyze the resulting sequences with the BLAST program.*

Lab One

BACKGROUND

Classification of bacteria and the 16S rRNA gene

KEY POINTS

- In the past, a bacterial species was defined according to the characteristics of the cell. If the bacteria had a unique set of characteristics, then it was considered a separate species.
- There are many problems with using observable characteristics to classify bacteria.
- The classification of organisms is now often based on the base sequence of the gene for the RNA in the small subunit of the ribosome. This RNA has the same function in all organisms, and much of the sequence is highly conserved in all living things. However, there are also variable regions where mutations can accumulate over time.
- The degree of sequence identity in the variable regions of the gene reflects the relatedness of organisms.
- In bacteria, the RNA found in the small subunit of the ribosome is called 16S rRNA.
- Base sequence comparisons of the 16S rRNA gene have been useful in classifying bacteria to the level of the genus.
- There are some difficulties in assigning relatedness based on the base sequence of a single gene.

The term "species" is a familiar one when applied to plants and animals and seems only a matter of common sense. In fact, biologists use a number of different definitions, depending on their point of view. For example, ecologists often define species in terms of geographical distribution and reproductive isolation. Another definition is that a species is composed of individuals who have the same set of genes. In other words, a species has a common gene pool. Still others view a species as a collection of individuals who have shared the same evolutionary history. With this definition, the degree of relatedness between two species is determined by how much of this history is shared.

How do these definitions apply to bacteria? Classifying bacteria according to the number of shared characteristics, the **phenetic method** of classification, was used for many years. If bacteria share the same characteristics, they are the same species, while related species will have many characteristics in common and distantly related bacteria will show few common traits. There are many problems with this approach. For one thing, many of the properties of a microorganism are determined by growing it in pure culture under different conditions. The large majority of bacteria have not been cultured, and in fact, the existence of many, previously unknown microorganisms has been inferred from the detection of unique sequences of chromosomal DNA fragments. A second problem is whether all shared properties should have equal weight in determining whether bacteria are related. Bacteria with flagella have a complex nanomachine that rotates the flagella to provide motility. Should this receive equal weight with the ability to metabolize lactose, a relatively simple trait? A third problem is that the properties of bacteria can change dramatically with the environment (and this includes the presence of other bacteria), so we would need to compare properties in multiple environments. Finally, bacteria exchange DNA by mechanisms of **horizontal gene transfer**, the passage of DNA from one cell to another, so that the properties of a microorganism can change suddenly (in evolutionary terms), often with dramatic results. This can lead to assigning a different species name to two variations of a single species that differ only by some horizontally transferred DNA.

The problems with the phenetic approach can be illustrated by two examples. *Bacillus anthracis*, *Bacillus cereus*, and *Bacillus thuringiensis* are regarded as separate species, as you can tell from the names. *B. anthracis* causes anthrax, an often fatal disease of livestock and humans. The species *B. cereus* is frequently found in the soil and can cause food poisoning. *B. thuringiensis* has the interesting property of producing a widely used crystalline insecticide, while its ability to cause disease in humans is exceedingly rare (Ibrahim et al., 2010). Each species is dramatically different, yet the differences are determined by horizontally transferred plasmids (extrachromosomal elements of DNA). *B. anthracis* has one set of plasmids that make it a serious pathogen. Isolates of *B. cereus* that cause food poisoning have a different set of plasmids responsible for the production of toxins such as cereulide, which causes vomiting. Other isolates, without these plasmids, are benign. Yet another set of plasmids is responsible for the production of insecticide by *B. thuringiensis*. Apart from the different plasmids, these three species are basically the same and should certainly be considered as one species (Rasko et al., 2005). The second example is the bacterium *Escherichia coli*, a single species. *E. coli* K-12 is in common use in microbiology laboratories and is benign, even if accidentally ingested. *E. coli* O157:H7 is a severe pathogen that causes bloody diarrhea and damage to the kidneys, sometimes resulting in permanent impairment or death. Because most of the virulent properties of *E. coli* O157:H7 are the result of horizontally acquired DNA and the chromosomes of the two are otherwise similar (Hayashi et al., 2001), each is considered to be a member of the same species. They are referred to as different **strains**, variants of a single species having different properties. As you can see, it is certainly difficult to maintain consistency with a purely phenetic approach.

The classification of organisms by evolutionary history is called the **phylogenetic method**. For decades, paleontologists have relied on fossils to identify and classify different species by this method. While bacterial fossils exist, they have few observable features and mostly testify to the antiquity of life on this planet rather than to evolutionary progression. Among the most dramatic bacterial fossils are stromatolites (Fig. 2-1), formed by the gradual accretion of sea sediment on the surface of

Figure 2-1 Stromatolites on the western coast of Australia (photographer Paul Harrison) CC BY-SA-3.0.

innumerable cells of cyanobacteria. Some stromatolytes are about 3.5 billion years old and represent the earliest fossil record of life on the planet.

All organisms contain RNA in the large and small subunits of the ribosome. In the 1970s, Carl Woese had the brilliant and revolutionary idea of classifying organisms by comparing the base sequences of the gene for the RNA in the small ribosomal subunit.

The method is based on the following:

1. A gene for the small-subunit ribosomal RNA is found in all organisms.
2. The function of the RNA is essential and the same in all organisms.
3. Regions of the gene can be mutated without loss of function.

The small-subunit rRNA maintains the structural integrity of the ribosome and is required for the initiation of translation at the correct site on the messenger RNA. Because these functions have remained unchanged, the base sequence of this gene has been substantially conserved over evolutionary history (Fig. 2-2). The conserved base pairs are probably essential for function. Mutations at these sites would be deleterious or lethal and would not be retained. However, there are also variable regions of the gene where mutations have little effect. These mutations are inherited

human	...GTGCCAGCAGCCGCGGTAATTCCAGCTCCAATAGCGTATATTAAAGTTGCTGCAGTTAAAAAG...
yeast	...GTGCCAGCAGCCGCGGTAATTCCAGCTCCAATAGCGTATATTAAAGTTGTTGCAGTTAAAAAG...
corn	...GTGCCAGCAGCCGCGGTAATTCCAGCTCCAATAGCGTATATTTAAGTTGTTGCAGTTAAAAAG...
Escherichia coli	...GTGCCAGCAGCCGCGGTAATACGGAGGGTGCAAGCGTTAATCGGAATTACTGGGCGTAAAGCG...
Synechococcus elongatus	...GTGCCAGCAGCCGCGGTAATACGGGAGAGGCAAGCGTTATCCGGAATTATTGGGCGTAAAGCG...

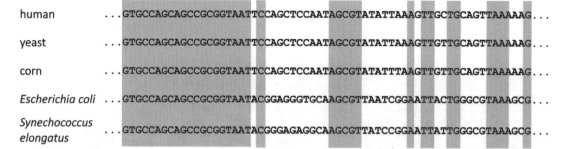

Figure 2-2 Different species and the similarity of rRNA gene sequences. Only part of the sequence is shown. *Synechococcus elongatus* is a member of the cyanobacteria, which are considered to be among the most ancient organisms on earth.

because they are neutral or nearly so: they do not affect the function of the RNA and therefore confer no significant disadvantage to the organism.

Some mutations result in the change of a single base pair in a DNA sequence (for example, GC → AT). To a good approximation, these mutations occur at a low but constant rate at each base pair. As a consequence, the number of base-pair differences in the variable regions of the small-subunit RNA gene is a measure of the relatedness of two organisms. If two different bacteria diverged recently (on an evolutionary timescale) from a common ancestor, the base sequence in the variable regions will be very similar. More distantly related bacteria, those that diverged a long time ago, will have had the time to accumulate a larger number of mutations. Because mutations occur randomly at different base pairs within the variable regions, the locations of these mutations will be different.

It is very important that the function of the small-subunit RNA has remained the same throughout evolution. Because of this, the target size, the number of base pairs where nonlethal mutations can occur, has been relatively constant. Sometimes genes acquire a new and different function during the course of evolution. In that case, the target size could change, with formerly nonessential base pairs becoming essential and vice versa. For example, if the target size became smaller (that is, a greater fraction of the mutations are now lethal or detrimental), then neutral mutations would be acquired at a slower rate. If the mutation rate was assumed to remain constant, this would lead to the conclusion that some organisms are more closely related than they really are.

In bacteria, the RNA in the small subunit of the ribosome is called **16S rRNA** (16S refers to the sedimentation rate during centrifugation). The 16S rRNA sequence can be used to identify many bacteria, generally to the level of the genus (i.e., *Bacillus*, *Salmonella*, etc.), but by itself is not the ultimate tool for classification. There is still disagreement about where to set the cutoff point (such as <99% base sequence similarity) in deciding whether two isolates are different species. In fact, there might be a good reason not to do so. Some groups of bacteria have nearly identical 16S sequences while overall genomic similarity, based on a technique called DNA-DNA hybridization, would indicate that the members of the group are different species (Janda and Abbott, 2007). In addition, there is evidence that the 16S rRNA gene itself can be horizontally transferred from species to species (Kitahara and Miyazaki, 2013). This can result in a sudden, large change in the 16S rRNA gene sequence. Also, the genome usually contains multiple copies of the 16S rRNA gene. This is not a problem if the sequence of each copy is the same, but species having two sets of genes with different sequences have been encountered. It is becoming apparent that other highly conserved genes, in addition to the 16S gene, should be examined in order to assess overall relatedness. This approach is called **multilocus sequence typing**. Now that complete DNA sequences for different bacteria are faster and easier to obtain, global methods of sequence comparison will be used more frequently and perhaps the definition of a bacterial species can finally be decided.

Polymerase chain reaction (PCR)

KEY POINT

- The polymerase chain reaction (PCR) is a method for amplifying the number of copies of a defined segment of DNA.

The **polymerase chain reaction (PCR)** has become a familiar tool in biology. Essentially, it is a method of increasing the number of copies of a defined segment of DNA. The amplified product simplifies cloning, sequencing, and other procedures. PCR will be outlined here using the amplification of the 16S rRNA gene as an example. Although you might know the basic idea of PCR, it is worthwhile to consider the procedure step-by-step.

DNA replication is catalyzed by enzymes called DNA polymerases. These enzymes require a template and a primer as well as deoxynucleoside triphosphates (dNTPs) (Fig. 2-3). DNA is synthesized by the stepwise addition of nucleotides to the 3'-OH end of the primer strand. The nucleotides incorporated at each step are determined by the template, according to the Watson-Crick base-pairing rules (G pairs with C, A pairs with T). PCR is similar in many ways to naturally occurring DNA replication, but there are a few key differences. In PCR, the primer is a short, chemically synthesized oligonucleotide with a sequence allowing it to hybridize (anneal) to the template DNA where the sequence is complementary. Two primers are used, one for each strand of the duplex DNA. Chromosomal DNA is the source of the template encoding the 16S gene (Fig. 2-4). First, the cells are broken open to expose the DNA. During this process, the chromosome does not remain intact but is sheared to a collection of smaller fragments. A small fraction of these fragments will contain the complete 16S gene, but because they are generated by forces that

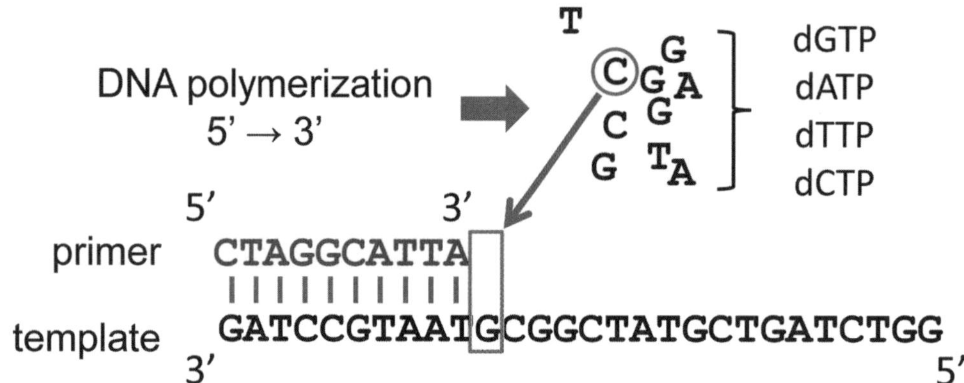

Figure 2-3 Primer annealed to a single DNA strand and the initiation of DNA synthesis.

randomly break the DNA, no two fragments containing the gene are likely to be exactly the same. The primers 27F and 1492R hybridize to the DNA at the beginning and end of the 16S gene (Fig. 2-4) so that an approximately 1,500-base-pair DNA fragment encoding the gene is amplified. These primers hybridize to conserved regions of the 16S gene so they can be used with many different bacteria, although it is unlikely there will be exact complementarity (Weisburg et al., 1991).

Figure 2-4 Overall scheme for the amplification of the 16S gene by PCR. bp = base-pairs DNA.

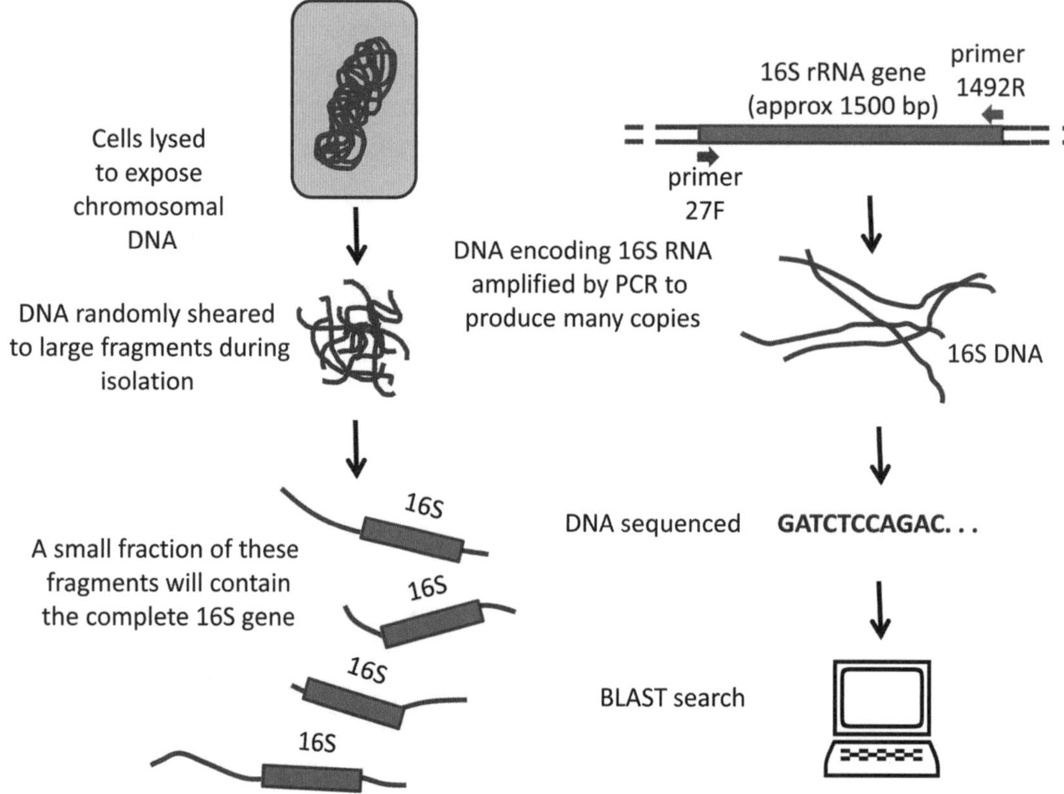

In the first step of the PCR procedure, chromosomal DNA, primers, and a heat-stable polymerase are added to a buffered solution containing the four dNTPs and additional components required for optimal synthesis. The primers must be present in vast molar excess compared to the DNA fragments containing the 16S gene. This requirement is explained below.

Cycle 1 (Fig. 2-5A): The mixture of chromosomal fragments, together with a heat-stable DNA polymerase and dNTPs, is heated to 95°C in buffer suitable for DNA synthesis. At this temperature, the DNA denatures and the two strands separate. The sample is then slowly cooled down. This gives time for complementary DNA strands to reanneal. Because there are so many primer molecules, each strand of DNA is likely to anneal with its complementary primer, to form the two products shown in the figure. At the same time, the chromosomal, single-stranded DNA will have difficulty finding complementary sequences among the many different chromosomal fragments that are present. The reaction is then brought to the optimum temperature for DNA synthesis. The DNA polymerase, which is able to withstand the high temperature required for the denaturation step, extends each primer to create new strands complementary to the template. During this process primers are consumed; that is, they are incorporated into the growing strand. The first cycle

Figure 2-5A PCR, first cycle (denaturation → annealing → polymerization).

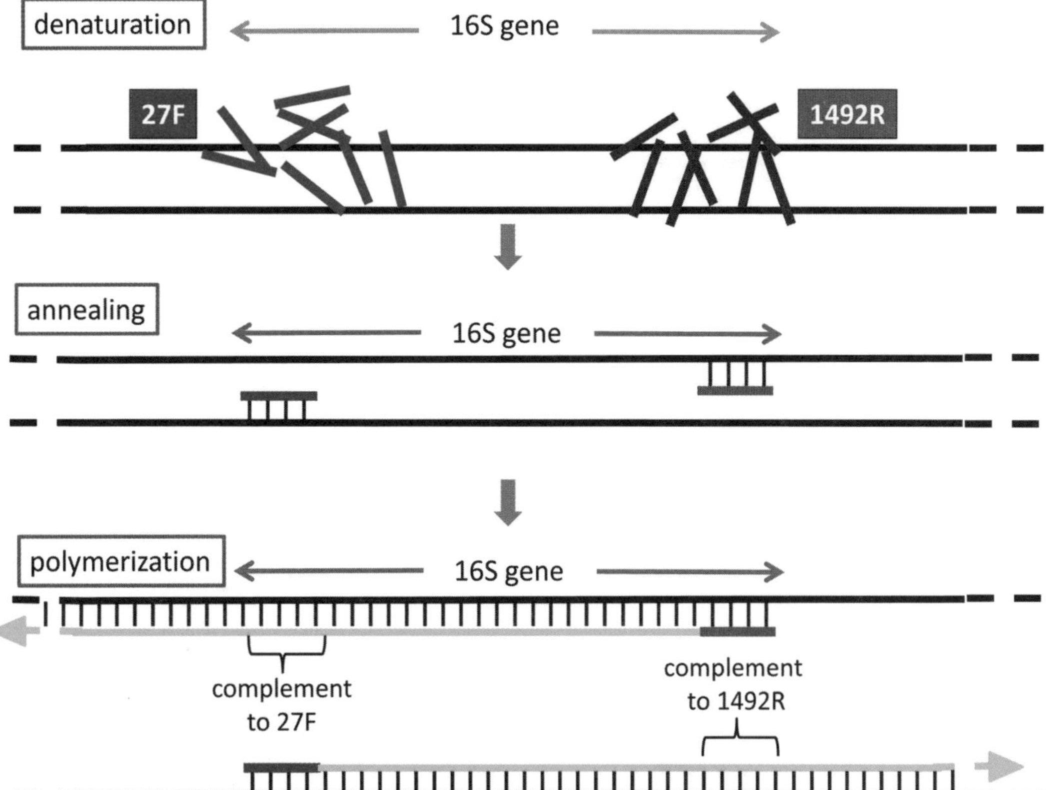

of synthesis (denature → anneal → polymerize) is now complete. The new DNA products have a defined 5' end, which is the 5' end of one of the primers. However, the 3' end is indeterminate: synthesis will generally extend to the end of the template, but these ends will all be different due to random shearing of the DNA during extraction. Notice also that the complement to each of the primers is present in the new strands, providing a new annealing site for each of the primers.

Cycle 2 (Fig. 2-5B): During this cycle, all the DNA is again denatured. After cooling, primers once again anneal to chromosomal DNA to yield the fragments described for cycle 1. However, the newly polymerized fragments can also act as templates for DNA synthesis since they contain sequences complementary to the primers (Fig. 2-5A). In this round of synthesis, DNA is synthesized from the 3' end of one primer to the 5' end of another. The result is a new DNA strand that extends the full length of the 16S rRNA gene.

Cycle 3 (Fig. 2-5C): In cycle 3, the newly synthesized DNA from cycle 2 can also serve as templates in the reaction. In this case, synthesis results in duplex DNA consisting of the 16S gene. This is exactly what we want, but at this point there

Figure 2-5B PCR, second cycle. Annealing and DNA synthesis are shown only for the DNA templates made in the first cycle.

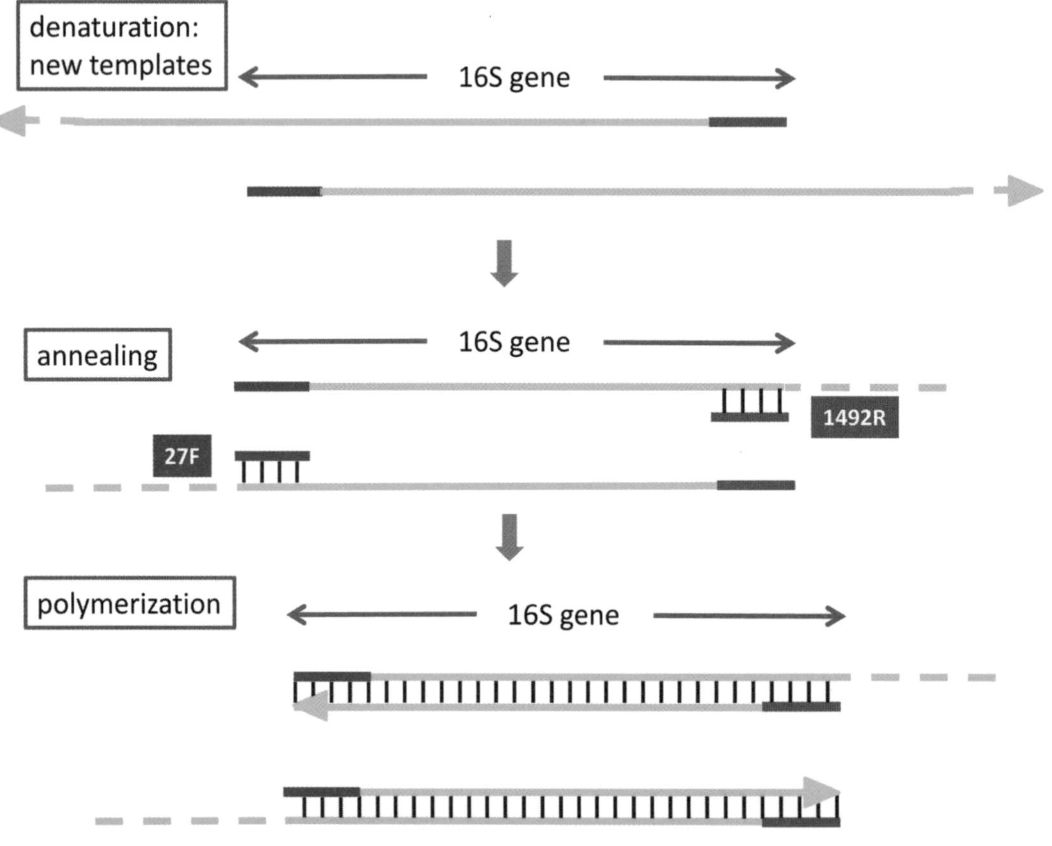

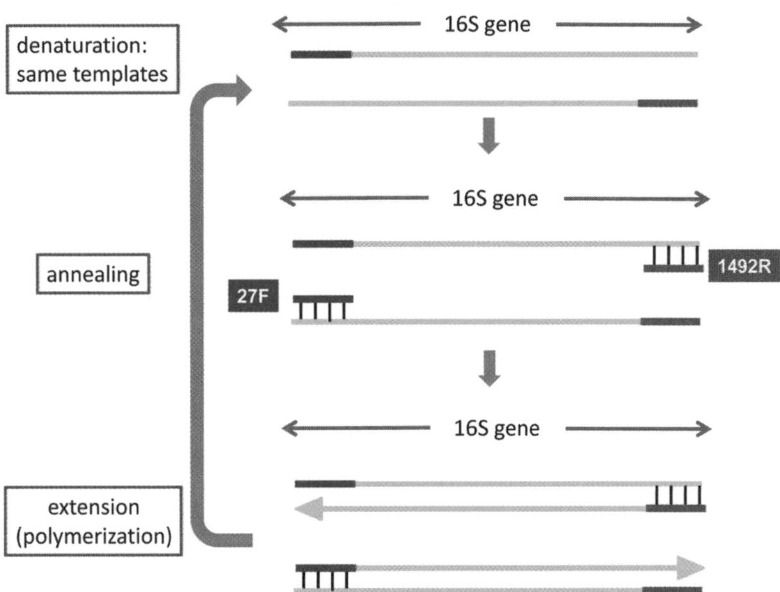

Figure 2-5C PCR, third cycle. Annealing and DNA synthesis are again shown only for novel DNA templates made in the previous cycle. The figure shows how the DNA synthesized in this cycle can be used as templates for the next and the following cycles, with more product of the same kind being generated.

has not been much synthesis overall and the chromosomal DNA fragments and the products of cycle 1 are still competing for primers as well. The key difference is that when the products of cycle 3 are used as the template during ensuing cycles, additional molecules of the same kind are generated. The number of these molecules increases geometrically in the reaction with every additional cycle, 2 → 4 → 8 → 16 and so on (Fig. 2-5D), because one molecule of (double-stranded) DNA provides two templates for synthesis during the next cycle. Soon, the number of these molecules is much greater than the input DNA or the products of the first and second cycles. However, in every cycle unincorporated primer is still required to initiate replication. A geometric increase in product (and therefore template for the next cycle) means there is a

Figure 2-5D Geometric increase in PCR product, due to the templates of one cycle resulting in product that creates more templates of the same kind for the next cycle.

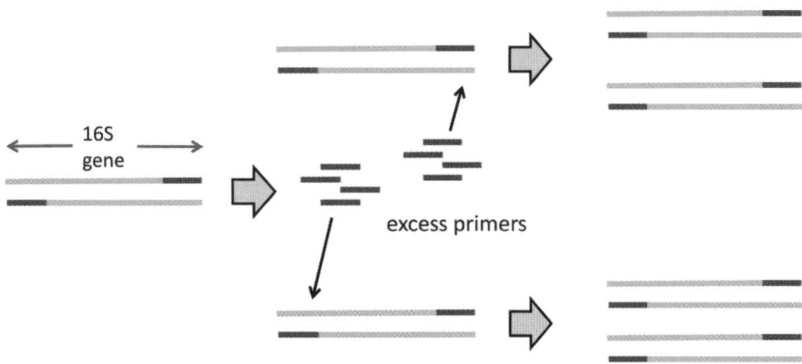

corresponding increase in the demand for primer and greater competition from the complementary strand during annealing. This is the major reason why a large number of primer molecules must be present at the outset.

Lab One

1. ***Use whole-cell PCR to amplify the 16S rRNA gene.***
2. *Verify by agarose gel electrophoresis that the PCR products are present.*
3. *Submit the PCR products for automated DNA sequencing at a core facility.*
4. *Analyze the resulting sequences with the BLAST program.*

Learning outcomes

After this lab, students will be able to:

a. Calculate volumes of stock reagents needed to set up a PCR reaction.
b. Dilute stock reagents to set up a PCR reaction.

I. Obtain enough DNA for sequencing: amplify the 16S rRNA gene by PCR

PROCEDURE

Setting up a PCR reaction

Each group should have freshly isolated colonies of each organism selected for PCR. You will be setting up 50-µl reactions consisting of the two primers 27F and 1492R; distilled, sterile water; and a master mix, which contains buffer, thermostable DNA polymerases, and the four dNTPs. The master mix and the primers will be provided as concentrated stocks:

	Stock concentration	Final concentration
Master mix	2×	1×
Primer 27F	2.5 µM	0.2 µM
Primer 1492R	2.5 µM	0.2 µM

Determine the needed volume of each reagent and complete the form below. Confirm your answers with the TA or instructor before proceeding. If you are unclear about how to proceed, read the end of this section. The volume of water is calculated by adding the volumes of all the other components and then subtracting from 50 µl.

Reagent	Volume (µl)
dH_2O	________
Primer 27F	________
Primer 1492R	________
2× master mix	________
Total volume	50 µl

1. Have ready the reaction components and two PCR tubes. Be sure each PCR tube is clearly labeled.

   The tubes are thin-walled to allow efficient heat transfer, but they are also fragile.

2. Add the dH_2O and both primers to the PCR tubes, then the master mix.

   The master mix is stored frozen: thaw at room temperature and then place on ice. Invert the tube several times to mix before using. Enzymes can lose activity if left at room temperature too long; thaw the master mix only when it is needed and add it to the PCR tube last.

3. Make sure that all the components of the reaction are mixed together at the bottom of the tube. If there are droplets on the side of the tube, spin the tube for a few seconds in a small tabletop microcentrifuge if available (rotor must be balanced) or lightly tap the tube on your bench.

   During subsequent handling, droplets are sometimes splashed onto the side of the tube. It is a good practice to look at the tube just before the PCR reaction and centrifuge again if necessary.

4. Using a sterile (flamed) inoculating *needle* (*not* a loop), transfer cells from a colony from each isolate into the solution in one of the tubes. Transfer a nearly invisible amount of cells: if you transfer a larger amount, the subsequent amplification might be inhibited.

5. Keep the PCR tubes on ice until everyone is ready to start the PCR.

The conditions for thermocycling will be preprogrammed into the machine and are shown in Table 2-1. The PCR should take about 3 hours.

6. At the end of thermocycling, the tubes will be cooled to 4°C. Once they have reached this temperature, you can take out your samples and store in the designated rack in the freezer (–20°C) for next week. Confirm that you can identify the labeling on your tubes.

Table 2-1 PCR Amplification of 16S rRNA Gene.

1. 94 °C / 2 min	Denaturation (strand separation of chromosomal DNA)
2. 94 °C / 20 sec	Denaturation at beginning of each cycle
3. 45 °C / 30 sec	Annealing primers and template
4. 68 °C / 1 min 30 sec	Polymerization
5. Go to step 2, 34 more times	Number of cycles (steps 2–4)
6. 68 °C / 5 min	Completion of any partially synthesized strands
7. 4 °C	Holding temperature

PROCEDURE

Diluting from stock solutions: using the formula $C_1 \times V_1 = C_2 \times V_2$

Microbiologists frequently use concentrated stocks of reagents and dilute these to the required final concentration. For example, suppose you have a stock solution containing 50 mM ATP and want to set up a reaction where the ATP concentration is 2 mM and the final volume is 50 µl. Using the formula above:

(**C**oncentration of ATP in stock solution) × (**V**olume of stock solution)
= (**C**oncentration of ATP in final reaction) × (**V**olume of reaction)

To solve for the volume of stock solution:

$$(50 \text{ mM ATP}) \times V_2 = (2 \text{ mM}) \times (50 \text{ µl})$$

$$V_2 = (2 \times 50)/50 = 2 \text{ µl}$$

It is helpful to keep in mind what the equation says: when you make a dilution, the amount (micrograms, millimoles, etc.) of the substance removed from the stock solution and the amount of this substance after dilution is the same. In both cases, this amount is the concentration times the volume. Just make sure that the units for concentration and volume are identical on both sides of the equation.

QUESTIONS

Questions are designated B1 to B6 according to the six levels of Bloom's taxonomy.

1. Will chromosomal DNA fragments containing only part of the 16S RNA gene be amplified by PCR? **(B2)**
2. You are trying to identify an unknown bacterium, so you set up a 50-µl PCR reaction using 2× master mix and 27F and 1492R primers. As you are putting away your stock tubes, you realize that by mistake you thought the concentration of the master mix was 20× instead of 2×. How will this change the final concentration of dNTPs and polymerase in your tube? **(B3)**
3. By mistake you set the PCR program to 17 cycles instead of the 35 cycles recommended in the manual. Will you get half as much product as a result? Explain. **(B3)**
4. The Taq polymerase used for the PCR synthesizes DNA at the rate of 1,000 base pairs per minute. What step in the PCR requires that we know this rate? Is the time allotted in the PCR program reasonable? Another thermostable polymerase, isolated from *Thermococcus kodakarensis*, an organism discovered in a volcanic vent, synthesizes DNA at the rate of approximately 125 base pairs per *second*. Explain how you would change the PCR program if you used this polymerase instead of Taq. **(B5)**

Lab Two

BACKGROUND

Agarose gel electrophoresis

KEY POINTS

- Agarose is a highly purified form of agar that forms a gel, a three-dimensional molecular mesh, when heated in liquid and cooled.
- During electrophoresis, DNA migrates through the mesh toward the positive terminal.
- When differently sized linear DNA fragments are compared, the shorter the fragment, the more quickly it migrates through the gel.
- DNA in the gel is visualized by staining with ethidium bromide and illuminating with UV light.

Agarose gel electrophoresis is one of the most important and frequently used techniques in molecular microbiology. Agarose is a highly purified form of agar, which is derived from red algae, a kind of seaweed. Agar consists mostly of two polysaccharide components, agarose, with low net charge, and agaropectin, which is highly charged. Charged groups interfere with electrophoresis, so agaropectin must be removed. The remaining agarose may have some other charged molecules and is frequently purified further to yield a nearly neutral product. When heated in liquid to boiling, the long agarose strands go into solution. As the liquid solution cools, the strands start to interact by hydrogen bonding to form a molecular network or three-dimensional mesh. This interaction results in a semisolid gel once the solution cools to about 35°C. During electrophoresis, negatively charged DNA migrates toward the positive terminal (anode). If this DNA is a mixture of linear DNA molecules of different sizes, the molecules will be separated by the agarose mesh according to their size, with the smallest molecules migrating most rapidly. Separation is based on the fact that the DNA molecules move in a way known as reptation, with each strand waving back and forth perpendicular to the direction of migration. Longer strands move more slowly because the greater amplitude of the waving strand makes it more likely that they will get caught on the strands of the mesh.

The most common arrangement is to have a slab of agarose with wells in the slab at one end (Fig. 2-6). The slab is submerged horizontally in a gel box containing buffer able to carry a current and connected to a power supply so that the positive terminal is at the far end. When the power is turned on, the negatively charged DNA migrates through the gel to the positive terminal, with the fragments of each size

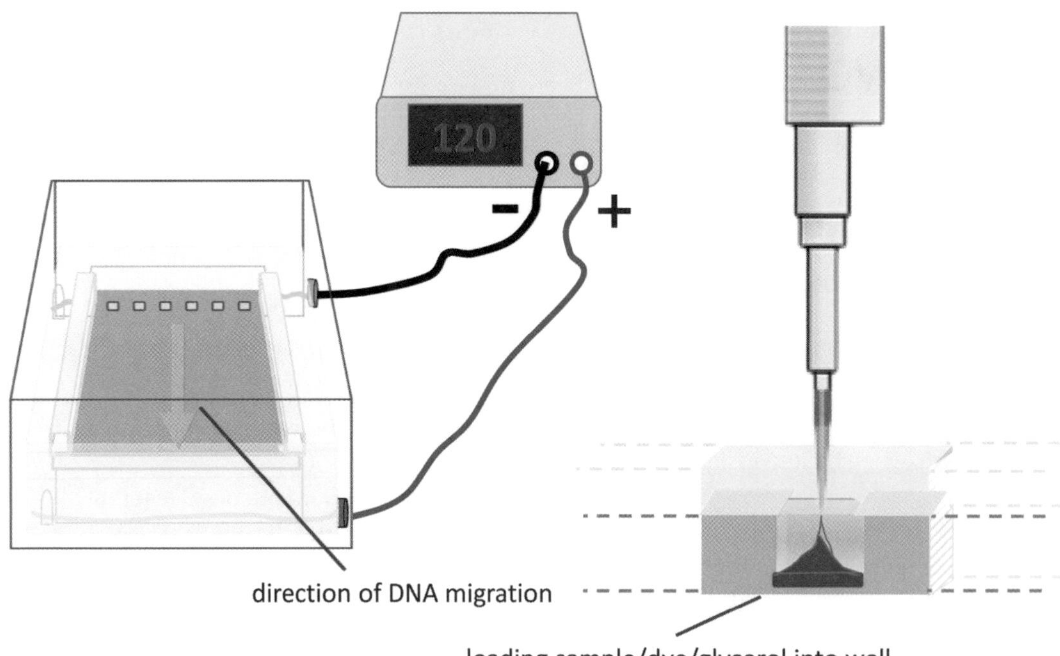

Figure 2-6 (Left) Design of typical apparatus for horizontal gel electrophoresis. (Right) Sample being loaded into well. Note tip is just above well opening. Sample contains glycerol and is denser than the buffer. Therefore it settles to the bottom of the well.

moving at a particular rate. Those fragments with the same size, and therefore the same rate of migration, move together as a DNA "band."

A mixture of glycerol (or some other dense component) and one or more visible dyes is added to the samples prior to electrophoresis. This mixture increases the density of the sample, allowing it to settle to the bottom of the well. One of the dyes migrates ahead of the DNA. The position of the dye band allows the progress of the electrophoresis to be monitored and helps to avoid running bands off the gel.

To visualize the DNA, it must first be stained. A common stain is ethidium bromide, which binds between the DNA base pairs. The DNA-ethidium bromide complex is highly fluorescent under UV illumination. Ethidium bromide can be added to the agarose gel before solidification or to the running buffer. Alternatively, the gel can be submerged in an ethidium bromide-containing solution after electrophoresis is complete. Ethidium bromide is hazardous (see below), and alternatives to ethidium bromide are available commercially. These are less toxic, but they are expensive and usually less sensitive.

A photograph of a sample gel, stained with ethidium bromide, is shown in Fig. 2-7. The DNA from four different bacteria was amplified by the class procedure. Notice that the PCR product of the 16S gene is about the same size on the gel, although the bacteria are not closely related.

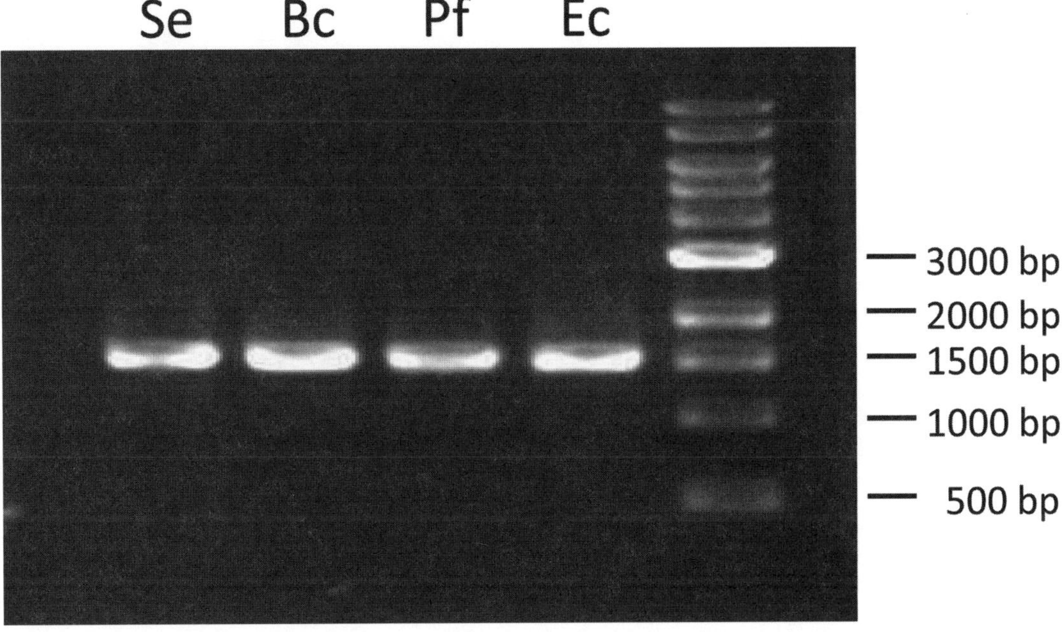

Figure 2-7 Agarose gel electrophoresis of 16S PCR product for *Staphylococcus aureus* (Se), *Bacillus cereus* (Bc), *Pseudomonas fluorescens* (Pf) and *Escherichia coli* (Ec). The marker DNA fragments and their sizes are on the right.

Dideoxy DNA sequencing

KEY POINTS

- DNA polymerization requires that each incorporated nucleotide has an OH group at the 3′ position.
- In dideoxy sequencing, a fraction of the polymerizing DNA is terminated at each nucleotide position due to the incorporation of a nucleotide containing H instead of OH at the 3′ position.
- Termination at each nucleotide position results in a set of fragments differing by one nucleotide. These fragments can be separated by electrophoresis.
- The DNA sequence can be determined from the size of each fragment and the identity of the terminating nucleotide.

DNA polymerases catalyze the stepwise addition of deoxynucleotides to the 3'-OH end of a DNA strand, according to the Watson-Crick base-pairing rules (Fig. 2-3). In this reaction, the 3'-OH of the growing strand attacks the α-phosphate of the entering deoxynucleoside triphosphate (dGTP, dATP, dCTP, or dTTP) and displaces the β- and γ-phosphates to form a new phosphodiester bond (Fig. 2-8, left). The reaction depends absolutely on the 3'-OH at the end of the DNA strand: if this is not present, then no further growth of the DNA strand can occur (Fig. 2-8, right).

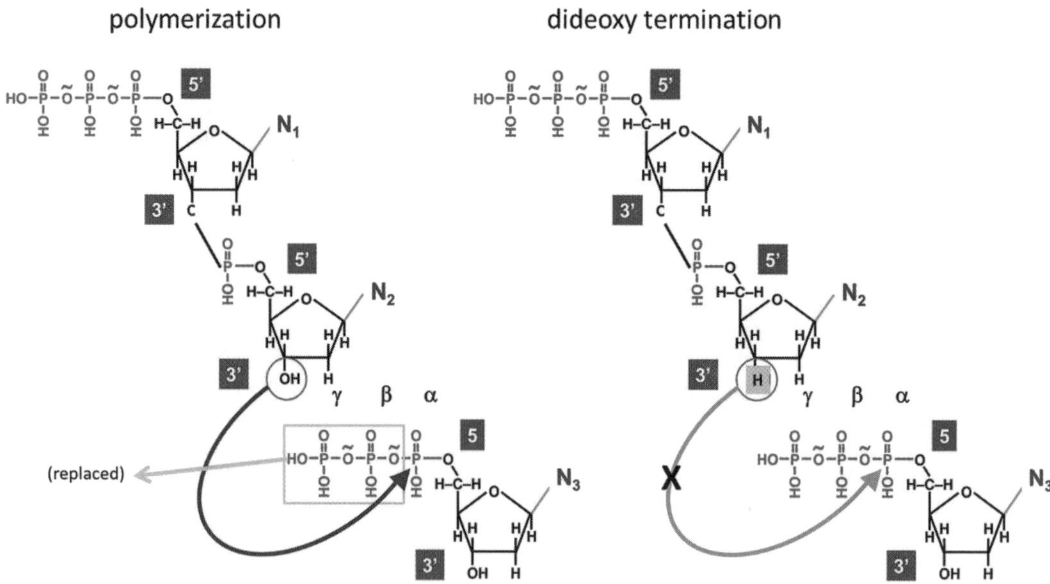

Figure 2-8 (Left) normal polymerization of DNA by formation of a new phosphodiester bond. (Right) polymerization failure (chain termination) due to absence of 3'—OH.

Dependence on the 3'-OH for polymcrization is the basis for the **dideoxy method of DNA sequencing.** If the OH is replaced with an H, phosphodiester bond formation does not occur and the extension of the chain is halted. In the original design, four separate reactions were set up (Fig. 2-9), each containing the template to be sequenced, a primer, the four dNTPs, and DNA polymerase. In addition, a different dideoxy (dd) NTP (ddGTP, ddATP, ddTTP, or ddCTP) was added to each reaction. What happens now in each reaction? As an example, consider reaction I (dGTP

Figure 2-9 The four reactions of dideoxy DNA sequencing.

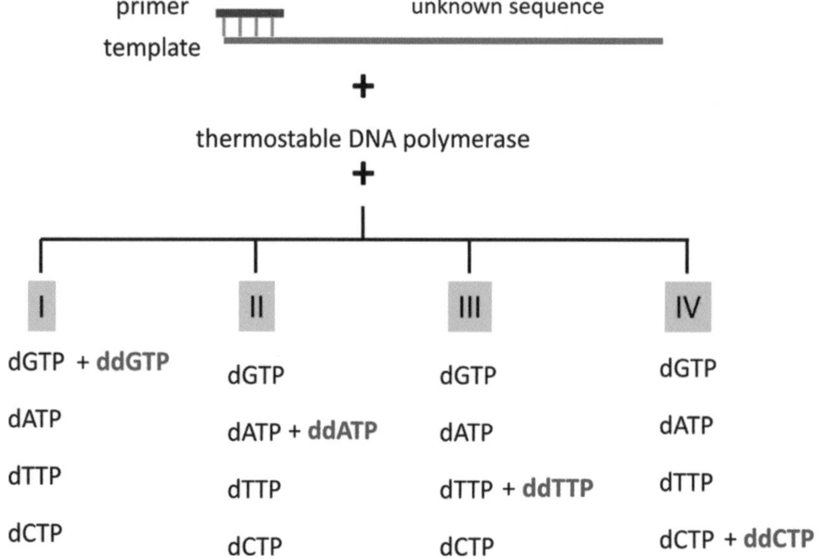

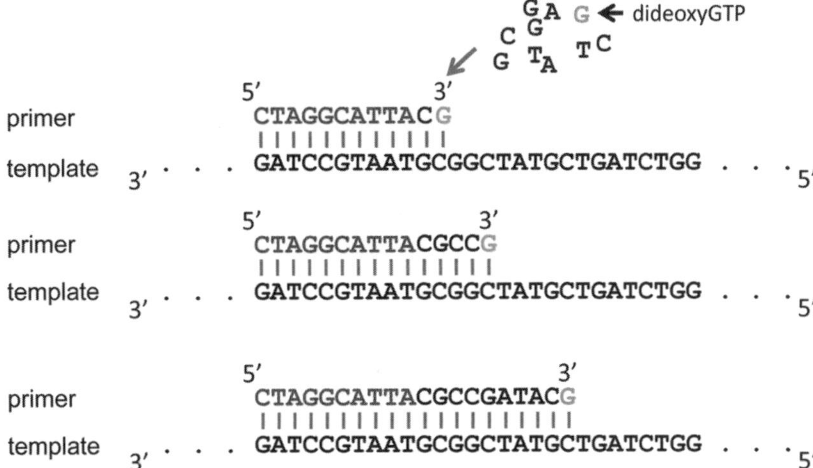

Figure 2-10 Products formed when DNA polymerization is terminated by dideoxy GTP.

and ddGTP) and the primer-template pair shown in Fig. 2-3. In this reaction (Fig. 2-10), the first base added during polymerization is dC. The second base added is a dG, but the reaction mixture contains both dGTP and ddGTP, so either can be incorporated. If dGTP is added, strand extension continues, but if ddGTP is added, there is no further extension of the DNA because of the missing 3'-OH. As a result, a DNA fragment is formed consisting of 12 bases (10 contributed by the primer). For the strands that incorporated dGTP instead of ddGTP, synthesis continues until the base-pairing rules direct the insertion of another dG. Once again, if ddGTP is added, extension of the strand is terminated, resulting in a fragment of 15 bases. If dGTP is added instead, then polymerization continues. Since there are a great number of primer-template molecules in the reaction, what will finally be generated is a set of fragments with different sizes and ending with ddG. In reaction II, containing ddATP (Fig. 2-9), all the fragments will end with ddA. However, the population of these fragments will have different sizes than those generated by termination with ddGTP. In fact, each one of the four reactions will generate a set of fragments having sizes that are not found in any of the other reactions. You can verify this yourself by determining the fragment sizes that would be terminated by ddCTP and ddTTP in Fig. 2-10. Taken together, the fragments from the four reactions make up a "ladder" of fragments differing by one nucleotide.

The DNA in each reaction is then denatured to separate the strands, and the single-stranded, dideoxy-terminated fragments separated by electrophoresis. These small fragments cannot be separated by agarose gel electrophoresis because the size of the agarose mesh is too large, resulting in all the fragments passing through the gel at the same rate. Instead, gels are made from cross-linked polyacrylamide. In these gels, small, single-stranded DNA fragments differing by only one nucleotide can be separated during electrophoresis, and this is exactly what is needed.

The fragments for each of the four reactions are shown schematically in Fig. 2-11 (left). The first incorporated base where dideoxy termination is possible is a C. This then generates the shortest fragment, and it will appear in the ddC lane. The next base is a G: termination here will result in the next-shortest fragment, which will be in the ddG lane. Continuing with termination at each succeeding base, the pattern of fragments shown in Fig. 2-11 (left) will be created. Every sequence will generate its own pattern of fragments, and from this pattern we can work backward to determine an unknown sequence. In the example here, suppose the sequence of the synthesized DNA (CGCCGA . . .) is unknown. From the bands, we can determine the sequence. The smallest band appears in the ddC lane, so the first base incorporated must be C. The second-smallest band is in the G lane, so the next base incorporated is a G. Working up the set of bands in order of size, we are able to derive the sequence. An actual gel is shown in Fig. 2-11 (right). The synthesized DNA was radioactively labeled and exposed to X-ray film for visualization. The sequence is not the same as the one on the left: can you read the sequence?

Modern dideoxy sequencing, carried out in "core facilities" of major universities and companies, incorporates a modification of the original procedure to increase automation and speed. The DNA synthesis step is done in a single tube, with all four ddNTPs present. As a result, termination is possible at every position and a ladder of bands is generated that increase in size by steps of one base. For each band, the dideoxy base that terminated the extension is determined by using ddNTPs each

Figure 2-11 (Left) Schematic representation of the migration of different dideoxy-terminated fragments after electrophoresis through a gel. The fragment synthesized in each case is in the orange box with an arrow drawn to the corresponding gel fragment. (Right) Real data for a different DNA sequence. Bands were radioactive and were visualized by using X-ray film.

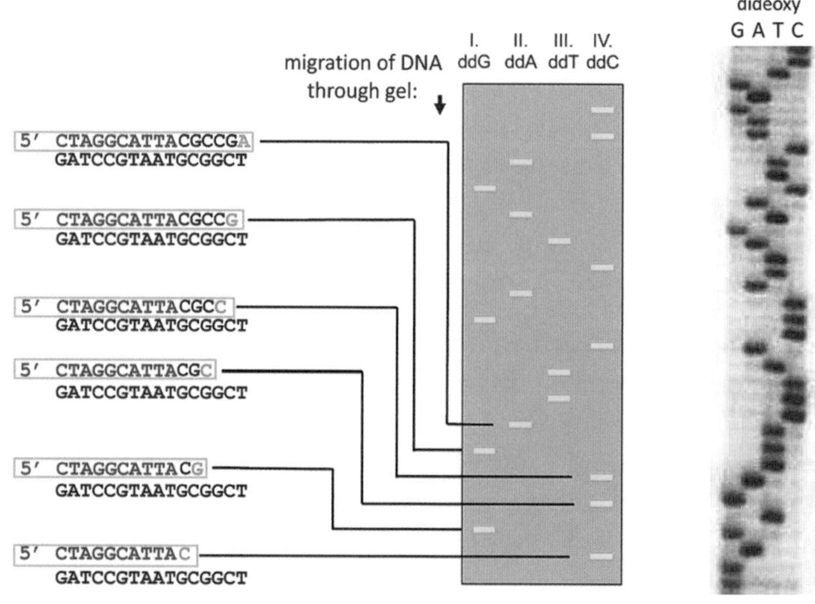

linked with a different fluorescent chemical. Because each of these fluoresces at a different wavelength, the color of the band identifies which ddNTP has been incorporated.

Instead of large slab gels, a capillary tube is used for electrophoresis (Fig. 2-12). The reaction products are applied at one end, and the bands separate as they migrate down the tube. At the other end, there is a window in the capillary. This allows a laser to shine on each band as it passes by. The color of each is recorded and displayed graphically as a series of colored peaks. Finally, a computer program calls out the base corresponding to each peak.

Lab Two

1. *Use whole-cell PCR to amplify the 16S rRNA gene.*
2. ***Verify by agarose gel electrophoresis that the PCR products are present.***
3. ***Submit the PCR products for automated DNA sequencing at a core facility.***
4. *Analyze the resulting sequences with the BLAST program.*

Learning outcomes

After this lab, students will be able to:

- **a.** Calculate the amount of agarose needed for a % wt/vol agarose gel.
- **b.** Load and run an agarose gel.
- **c.** Interpret the results of gel electrophoresis.

Figure 2-12 Modern method for detecting dideoxy-terminated fragments. The graphical output (chromatogram) and the base "calls" are from an actual sequencing run. With permission of Dr. Robert Lyons, DNA Sequencing Core, University of Michigan.

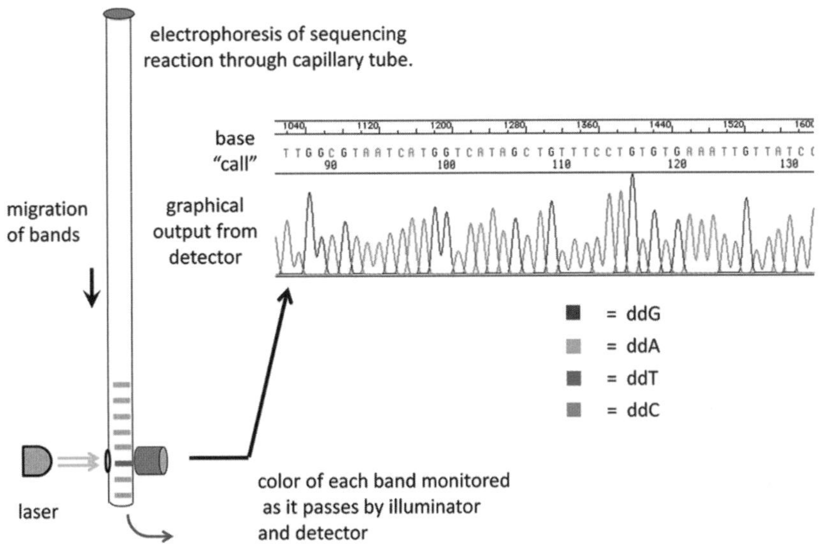

I. Visualize the PCR product by agarose gel electrophoresis

PROCEDURE

Making an agarose gel and carrying out gel electrophoresis

SAFETY Ethidium bromide is carcinogenic and mutagenic. Always wear gloves when handling stained gels and solutions containing ethidium bromide. Dispose of solutions containing ethidium bromide properly as described by the instructor.

The Tris-borate buffer used in electrophoresis is poisonous when ingested.

Obtain an agarose gel or prepare a gel in the lab. The size of the gel and the method for forming the gel will depend on the electrophoresis apparatus. If preparing a gel, keep in mind the following.

1. Weigh out the correct amount of agarose powder. For a 0.8% (wt/vol) gel, this means 0.8 g of agarose per 100 ml gel volume. For example, if the gel volume is 30 ml, you will need 0.24 g of agarose. Your instructor will tell you the volume of the gel for the apparatus you are using.
2. Add the powder to a volume of electrophoresis buffer (*not* water) that is equal to the gel volume. For example, If the gel volume is 30 ml, then use 30 ml of the buffer. Have the buffer in a microwaveable container such as an Erlenmeyer flask. The powder will not go into solution.

   Use a container that is sufficiently large to avoid boil-over during microwaving. A 125-ml flask is suitable for 30 ml of liquid. Do not tightly cap the container. Covering with a loose cap will minimize loss of liquid.
3. Microwave until the agarose powder has dissolved. The agarose will not go into solution until the liquid is about 100°C. After microwaving, let cool for a few seconds and swirl gently. Examine to be sure that the solution is clear and uniform.
4. Let the agarose solution cool to about 55 to 60°C. Add one drop of ethidium bromide solution and swirl gently to mix before pouring the gel.

SAFETY Do not microwave agarose in a tightly sealed container. Container and contents will be very hot.

Details for setting up the gel electrophoresis unit will be provided by the instructor.

1. Thaw the PCR samples. On a piece of Parafilm place:

   2 µl PCR samples
   8 µl H_2O
   2 µl glycerol-tracking dye or equivalent
2. Mix by gently pipetting the liquid in and out of the tip several times. Avoid making air bubbles.

   Air bubbles in the tip can result in loss of part of the sample while loading, due to their sudden discharge from the tip.
3. Draw up one of the samples into the pipette tip.

   Use a micropipette set for 12 µl and make sure there is no air between the sample and the small tip opening. If there is, slowly reduce the volume until the air is expelled. Do this before placing the end of the tip in gel buffer.

4. Deposit the sample into one well of the agarose gel. Make sure that the pipette tip is in or just above the well but not touching the well bottom. Repeat for the second sample. To one lane in each gel, add 12 µl of the 1-kb ladder (1 kb = 1,000 base pairs). The ladder is a collection of DNA fragments mostly differing in size by 1 kb (Fig. 2-7).
5. Carry out electrophoresis at 100 V until the blue dye is about three-quarters of the way down the gel (about 45 to 60 minutes). Stain the gel and photograph under UV illumination. A sample result is shown in Fig. 2-7. Store your remaining PCR products at –20°C.

SAFETY Never look directly into UV light. Always wear protective eyewear or a UV shield when viewing gels under UV illumination.

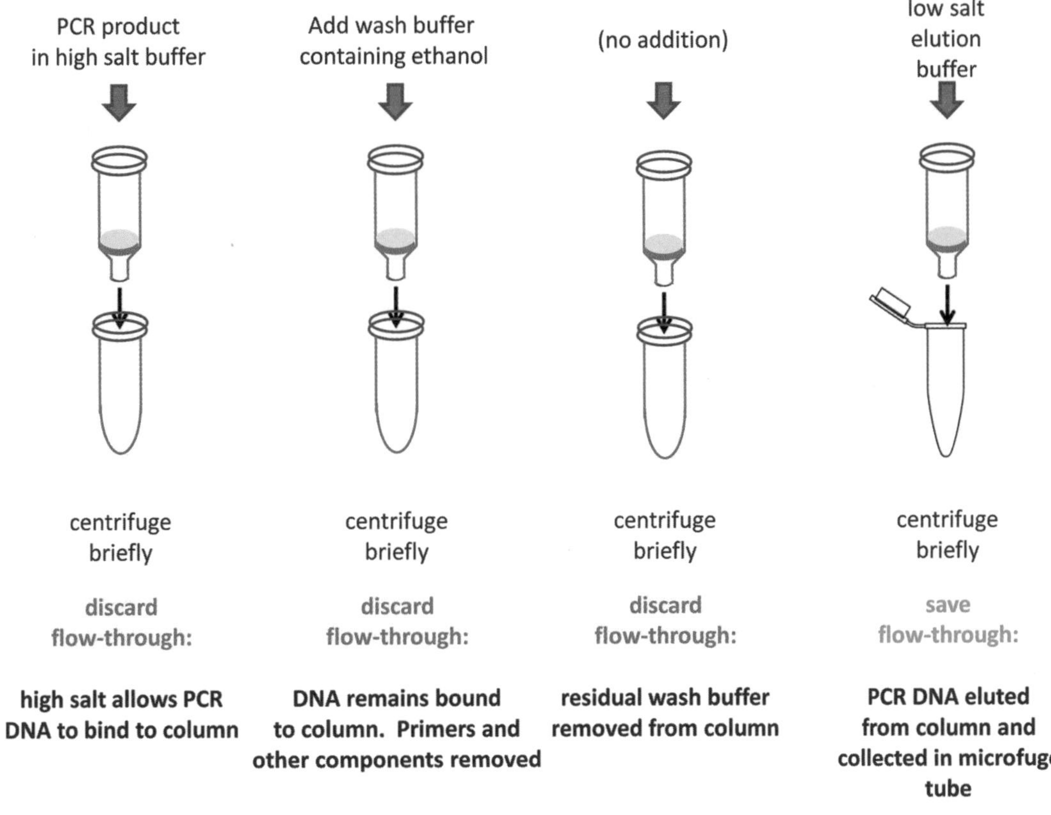

Figure 2-13 General method for cleaning PCR DNA. The PCR reaction mixture is first diluted in a high-salt buffer that disrupts base-pairing, allowing the DNA to bind to the column during centrifugation. The silica used in the column does not bind fragments smaller than approximately 100 bases, so primers as well as the other components are removed by the wash buffer. The wash buffer contains ethanol, which interferes with other applications. For this reason, residual buffer in the column is removed by an additional centrifugation. Finally, a low salt buffer is added to the column. This releases the DNA, which is then collected by centrifugation into a microfuge tube.

II. Submit samples for DNA sequencing

If a PCR was successful, sample the remaining reaction and add to a 1.5-ml tube. The volume of the sample will be indicated by your instructor. Make sure the tube is clearly labeled. Your instructor will give these samples to a sequencing facility. One of the two PCR primers, 27F or 1492R, will be used to prime DNA synthesis in the reaction.

Before sequencing, your PCR product will be "cleaned." The reaction mixture contains unused primers, dNTPs, Taq polymerase, and buffer components required for the reaction, all of which will be removed prior to the sequencing step. Alternatively, the instructor will clean your samples or ask you to do it. There are a number of different methods for cleaning a PCR reaction, but the simplest (and most popular) is to use a spin column containing silica. A common, general method is outlined in Fig. 2-13.

QUESTIONS

1. Many times, both strands of the same fragment of DNA are sequenced. If the sequence is correct, then the sequence for one strand should be the complement of the sequence for the other.
   a. You decide to sequence the complementary strand in Fig. 2-9. In the figure, what would be a good location for the primer for this reaction? (B3)
   b. Would this primer be complementary to the strand shown in the figure? (B3)
   c. To sequence your PCR product, 27F or 1492R will be used as the primer. Why can either primer be used? What is the relationship between the sequences obtained in each case? (B3)
2. In preparing your samples for gel electrophoresis, you added 2 µl of the dye-glycerol solution to your diluted PCR sample for a final concentration of 1× dye-glycerol. How would you describe the starting concentration of the dye-glycerol solution? (1×, 2×, 3×, etc.)? (B3)
3. You need 10 ml of the dye-glycerol solution for your students and decide to use the recipe below. You have on hand 100% glycerol and powdered bromophenol blue. How would you make this solution? (B3)

   Glycerol (30% vol/vol)

   Bromophenol blue (0.25% wt/vol)

   Add distilled H_2O to a final volume of 10 ml

   "Wt/vol" means the weight of a substance dissolved and then diluted until the desired concentration is reached. Usually the weight is given in grams, the volume in milliliters, and the concentration expressed as a percentage. For example: suppose you want 500 ml of a 5% solution (wt/vol) of sodium chloride in water. Dissolve 25 g sodium chloride in water; make up to a final volume of 500 ml. 25 g/500 ml = 5 g/100 ml = 5% (wt/vol) sodium chloride solution.

   "Vol/vol" means the volume of a substance diluted to a *final volume* so that the desired concentration is reached. Usually the concentration is given as a percentage. For example: suppose you want 200 ml of a 15% (vol/vol) solution of glycerol in water. To 30 ml of 100% glycerol add water to a final volume of 200 ml. 30 ml/200 ml = 15 ml/100 ml = 15%.

4. 100% glycerol is very viscous and difficult to pipette accurately. Instead, you decide to use a 50% (vol/vol) solution, which is much easier to handle. How would the calculations in question 4 change? Note that this is a $C_1V_1 = C_2V_2$ problem (Lab One). (B3)

Solving Challenge Two

1. Use whole-cell PCR to amplify the 16S rRNA gene.

2. Verify by agarose gel electrophoresis that the PCR products are present.

3. Submit the PCR products for automated DNA sequencing at a core facility.

4. Analyze the resulting sequences with the BLAST program.

BACKGROUND

BLAST (Basic Local Alignment Search Tool) is a suite of free bioinformatics programs that compare an unknown DNA sequence with those in the database and report close matches. It is provided by the National Center for Biotechnology Information (NCBI). If your sequence most closely matches those from organisms of a single genus (e.g., *Pseudomonas* or *Bacillus*), then your unknown probably belongs to that genus. Often it is not possible to identify the species from the 16S sequence alone. The programs are also used to compare the amino acid sequences of different proteins.

Learning outcomes

After this exercise, students will be able to:

a. Analyze sequencing results using BLAST to identify an unknown bacterium to the genus level.

Identifying the unknown microorganism from the 16S rRNA gene sequence

PROCEDURE

Preparing the sequence for analysis

When the sequencing result is returned, the data will look like that shown in Fig. 2-14. N's in the sequence indicate ambiguity about the correct base at that position. N's mostly occur at the beginning and end of a "read," and these can be cropped. A few N's elsewhere are not uncommon and are the reason both strands of a DNA fragment are often sequenced. Good-quality sequence contains at least 400 identified bases (G, A, T, and C) and a small proportion of N's among these bases. If there are many N's throughout, then there was too little DNA or there was some other problem. In this example, the bases in red were cropped and the remaining bases used for the BLAST search.

NNNNNNNNNNNNNNNNNTANNTCCTCCCGAAGGTTAGACTAGCTACTTCTGGTGCAACCCACTCCCATGGTGTGACGGG
CGGTGTGTACAAGGCCCGGGAACGTATTCACCGCGACATTCTGATTCGCGATTACTAGCGATTCCGACTTCACGCAGTCGA
GTTGCAGACTGCGATCCGGACTACGATCGGTTTTATGGGATTAGCTCCACCTCGCGGCTTGGCAACCCTTTGTACCGACCAT
TGTAGCACGTGTGTAGCCCAGGCCGTAAGGGCCATGATGACTTGACGTCATCCCCACCTTCCTCCGGTTTGTCACCGGCAG
TCTCCTTAGAGTGCCCACCATTACGTGCTGGTAACTAAGGACAAGGGTTGCGCTCGTTACGGGACTTAACCCAACATCTCAC
GACACGAGCTGACGACAGCCATGCAGCACCTGTCTCAATGTTCCCGAAGGCACCAATCCATCTCTGGAAAGTTCATTGGAT
GTCAAGGCCTGGTAAGGTTCTTCGCGTTGCTTCGAATTAAACCACATGCTCCACCGCTTGTGCGGGCCCCCGTCAATTCATT
TGAGTTTTAACCTTGCGGCCGTACTCCCCAGGCGGTCAACTTAATGCGTTAGCTGCGCCACTAAGAGCTCAAGGCTCCCAA
CGGCTAGTTGACATCGTTTACGGCGTGGACTACCAGGGTATCTAATCCTGTTTGCTCCCCACGCTTTCGCACCTCAGTGTCA
GTATCAGTCCAGGTGGTCGCCTTCGCCACTGGTGTTCCTTCCTATATCTACGCATTTCACCGCTACACAGGAAATTCCACCAC
CCTCTACCATACTCTAGCTTGCCAGTTTTGGATGCAGTTCCCAGGTTGAGCCCGGGGATTTCACATCCAACTTAACAAACCA
CCTACGCGCGCTTTACGCCCAGTAATTCCGANTAACGCTTGCACCCTCTGTATTACCGCGGCTGCTGGCACAGANTTAGNC
NGNNGCTTATTCTGTNGGTAACGTCAAANNNGCANAGTANTNNTNNNNNNNNNCNTNCTCNCNNCTTNAAAGNGCTTT
ANNANNCCNANANNTNCTTCANNNNCNCNGNANNNNNNGNATCNNNNTTCNNCNNNNNNNNNNNNCNCNNNNN
NNNNNTNNNNNNNNNNNNNNNNNNNNNNTCANNNNNNNNNNACNNNNNNNNNNNNNNNNNNNNNNNNNNNNN
NNNNNNNNNNNTTNNNNNNNNNNNNNN

Figure 2-14 Base sequence data returned from sequencing center.

PROCEDURE

Doing a BLAST search

Step one: In your browser, navigate to: http://blast.ncbi.nlm.nih.gov/Blast.cgi. The design of the opening and following screens changes periodically, but the overall structure will be similar.

A complicated page comes up (Fig. 2-15A), but much of it can be ignored. Click on "Nucleotide BLAST." This directs the program to the database of DNA sequences.

Step two (Fig. 2-15B)**:** Paste your sequence into the "Enter Query Sequence." Do not use the return key or punctuation marks.

Step three (Fig. 2-15C)**:** You will now tell the program what part of the nucleotide sequence database you would like to search. You are only interested in comparing the 16S sequences of bacteria, and it would be time-consuming and uninformative to use the entire database. Bacterial 16S sequences are frequently compared, and you can select the correct part of the database by simply scrolling down the database choices and selecting "16S ribosomal RNA sequences (Bacteria and Archaea)."

Click "BLAST" and wait. The screen will periodically refresh during this time. When the search is completed, the "BLAST Results" page will appear.

Step four (Fig. 2-15D)**:** Scroll down the results page. Sequences from the database that are similar to your sequence are listed. Each match is assigned a number of statistical values, including a score and an E value (the expect value). The score is a measure of the "quality" of the match between your sequence and the indicated sequence in the database. Points are added based on the number of aligned bases and the length of the alignment. Points are subtracted when there are gaps, regions of nonalignment within the match. This statistic depends on the size of the database: the larger the database, the more likely that the quality of the alignment, and therefore the score, is due to chance. It is easy to fool oneself: a match may have a high score and look significant, but is it? This is where the E value is useful. This statistic is a measure of the probability that the score for a match could have been the result of chance. The E value takes into account both the size of your sequence and the size of the database. An E value of 0 means that the probability of the match being due to chance is virtually zero, while an E value of 2 means that one would expect to find two matches with the same score (or better) occurring by chance. Very low E values indicate the matches are significant. The unknown organism in the example is *Pseudomonas*, but the species can't be absolutely determined from these results. Scroll down: the actual alignments are shown below the summary results ("Query" is your sequence).

Figure 2-15(A-D) Steps in a BLAST search. Images modified from US National Library of Medicine, National Center for Biotechnology Information website (https://blast.ncbi.nlm.nih.gov/Blast.cgi). With permission from the US National Library of Medicine, National Center for Biotechnology Information.

A

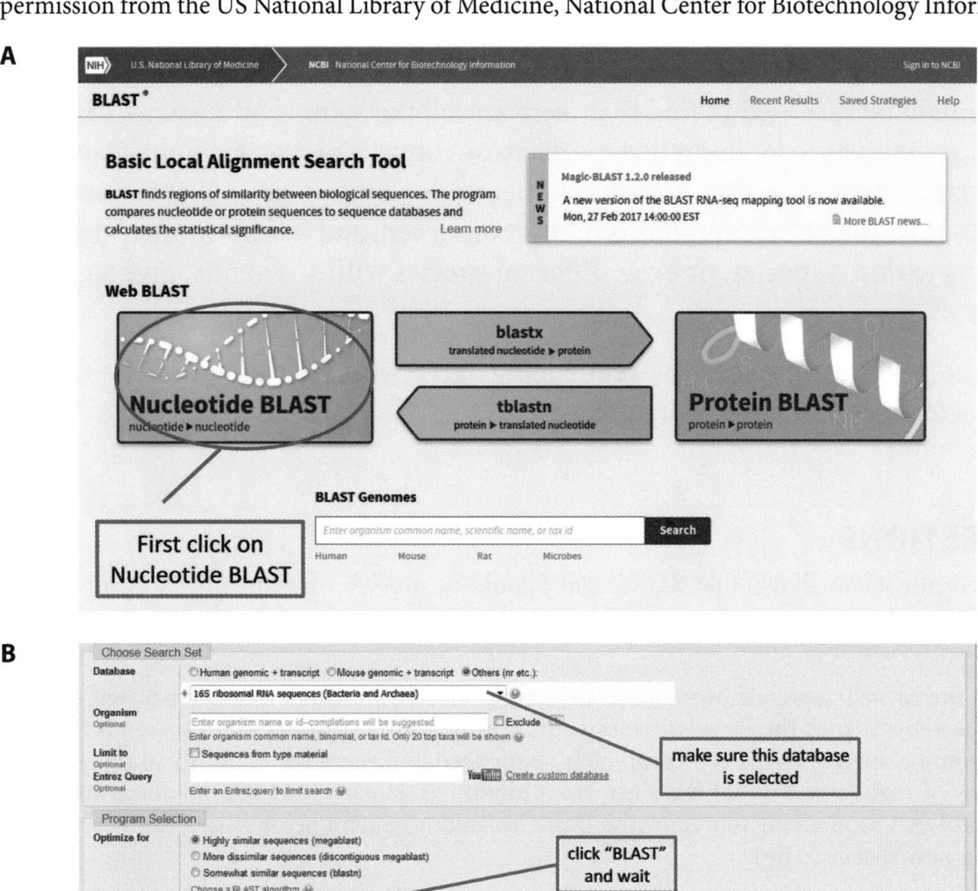

C

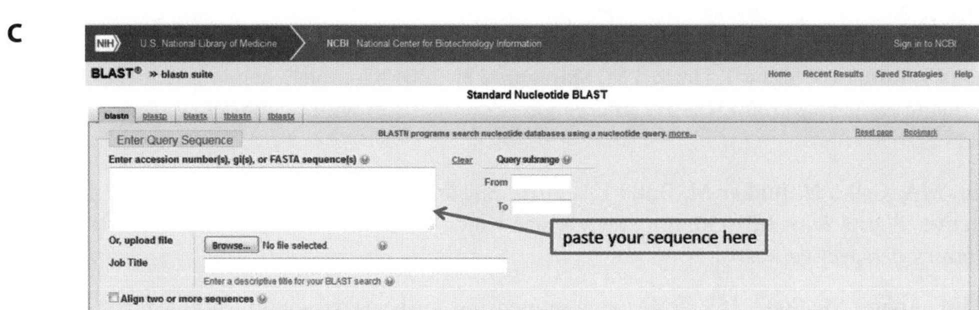

D

Scroll down to "Descriptions" to see list of organisms

Sequences producing significant alignments:

Select: All None Selected:0

Alignments Download GenBank Graphics Distance tree of results

Description	Max score	Total score	Query cover	E value	Ident	Accession
Pseudomonas aeruginosa strain SNP0614 16S ribosomal RNA gene, partial sequence	1164	1164	100%	0.0	99%	NR_118644.1
Pseudomonas aeruginosa PAO1 strain PAO1 16S ribosomal RNA, complete sequence	1164	1164	100%	0.0	99%	NR_074828.1
Pseudomonas aeruginosa strain DSM 50071 16S ribosomal RNA gene, partial sequence	1164	1164	100%	0.0	99%	NR_117678.1
Pseudomonas aeruginosa strain NBRC 12689 16S ribosomal RNA gene, partial sequence	1164	1164	100%	0.0	99%	NR_113599.1
Pseudomonas aeruginosa strain ATCC 10145 16S ribosomal RNA gene, partial sequence	1164	1164	100%	0.0	99%	NR_114471.1
Pseudomonas guezennei strain RA26 16S ribosomal RNA gene, complete sequence	1136	1136	100%	0.0	99%	NR_114957.1
Pseudomonas otitidis strain MCC10330 16S ribosomal RNA gene, complete sequence	1136	1136	100%	0.0	99%	NR_043289.1
Pseudomonas resinovorans NBRC 106553 strain NBRC 106553 16S ribosomal RNA, complete sequence	1131	1131	100%	0.0	98%	NR_103921.1
Pseudomonas indica strain NBRC 103045 16S ribosomal RNA gene, partial sequence	1120	1120	100%	0.0	98%	NR_114196.1
Pseudomonas resinovorans strain ATCC 14235 16S ribosomal RNA gene, complete sequence	1120	1120	100%	0.0	98%	NR_112062.1
Pseudomonas panipatensis strain Esp-1 16S ribosomal RNA gene, partial sequence	1114	1114	100%	0.0	98%	NR_044209.1
Pseudomonas indica strain IMT37 16S ribosomal RNA gene, partial sequence	1114	1114	100%	0.0	98%	NR_028801.1
Pseudomonas aeruginosa strain DSM 50071 16S ribosomal RNA gene, complete sequence	1110	1110	100%	0.0	98%	NR_026078.1
Pseudomonas citronellolis strain NBRC 103043 16S ribosomal RNA gene, partial sequence	1092	1092	100%	0.0	97%	NR_114194.1
Pseudomonas delhiensis strain RLD-1 16S ribosomal RNA gene, partial sequence	1092	1092	100%	0.0	97%	NR_043731.1

You might be surprised to find that there are a lot of significant matches to your sequence. One reason for this is that the database contains partial or complete sequences for many different strains, each a member of the same species. You can see this in Fig. 2-15D. The 16S rRNA gene is highly conserved, and it is no wonder that all the strains within a species, or even different species within a genus, give significant matches.

Do the genus assignments by sequencing agree with your assignments from the characteristics of the microorganism?

QUESTIONS

1. Recently it was shown that the 16S gene could be moved from one microorganism to another by horizontal gene transfer. How could this complicate using the 16S gene to determine relatedness? (B5)
2. In the BLAST search shown above, the results clearly indicated that the isolated organism was a member of the *Pseudomonas* genus. The genomes of about 10 of the common *Pseudomonas* species have been completely sequenced, but there are certainly many other species of *Pseudomonas* that have not been identified. How would you go about using PCR and DNA sequencing to determine if the isolated organism belongs to a known species or is a new species? (B6)

BIBLIOGRAPHY

Hayashi T, Makino K, Ohnishi M, Kurokawa K, Ishii K, Yokoyama K, Han CG, Ohtsubo E, Nakayama K, Murata T, Tanaka M, Tobe T, Iida T, Takami H, Honda T, Sasakawa C, Ogasawara N, Yasunaga T, Kuhara S, Shiba T, Hattori M, Shinagawa H. 2001. Complete genome sequence of enterohemorrhagic *Escherichia coli* O157:H7 and genomic comparison with a laboratory strain K-12. *DNA Res* **8:**11–22.

Ibrahim MA, Griko N, Junker M, Bulla LA. 2010. *Bacillus thuringiensis*: a genomics and proteomics perspective. *Bioeng Bugs* **1:**31–50. http://www.mendeley.com/catalog/bacillus-thuringiensis-genomics-proteomics-perspective/.

Janda JM, Abbott SL. 2007. 16S rRNA gene sequencing for bacterial identification in the diagnostic laboratory: pluses, perils, and pitfalls. *J Clin Microbiol* **45:**2761–2764. http://www.pubmedcentral.nih.gov/articlerender.fcgi?artid=2045242&tool=pmcentrez&rendertype=abstract.

Kitahara K, Miyazaki K. 2013. Revisiting bacterial phylogeny: natural and experimental evidence for horizontal gene transfer of 16S rRNA. *Mob Genet Elements* **3:**e24210. http://dx.doi.org/10.4161/mge.24210.

Rasko DA, Altherr MR, Han CS, Ravel J. 2005. Genomics of the *Bacillus cereus* group of organisms. *FEMS Microbiol Rev* **29:**303–329.

Weisburg WG, Barnes SM, Pelletier DA, Lane DJ. 1991. 16S ribosomal DNA amplification for phylogenetic study. *J Bacteriol* **173:**697–703.

challenge Three

Choosing an antibiotic to alleviate the symptoms of Crohn's disease

Crohn's disease is a chronic inflammation of a part of the gastrointestinal tract, most frequently at the end of the small intestine. The disease is probably due to a variety of factors including genetic predisposition and a hyperactive immune system. Individuals with the disease often produce antibodies against *Pseudomonas fluorescens*, and there is a correlation between the magnitude of the antibody response and the severity of the disease (Arnott et al., 2004). You have isolated a fluorescent *Pseudomonas* strain from a patient and want to determine if treatment with an antibiotic will lessen the symptoms of Crohn's disease. The first step will be to determine which antibiotics are active against the *Pseudomonas* strain and the required concentration of each.

QUESTIONS BEFORE YOU BEGIN THE CHALLENGE

1. Different antibiotics will be tested for their effectiveness against a growing culture of the *Pseudomonas* strain, and the results will be used to select an antibiotic for clinical trials. What assumptions are being made here?
2. How would you state your answer to question 1 in the form of a hypothesis?
3. How would you test the hypothesis formulated in question 2?
4. What ethical issue might arise in testing the hypothesis?

Strategy for Challenge Three

1. *Construct a graph of bacterial growth by measuring optical density.*
2. *Determine the concentration of viable cells by serial dilution and plating.*
3. *Calculate the relationship between optical density and viable counts for exponential-phase cells.*
4. *Determine the minimal inhibitory concentrations of different antibiotics by 2-fold dilutions in broth medium.*
5. *Interpret the results to identify the antibiotics that might be useful for treatment.*

Lab One

BACKGROUND

Exponential growth

KEY POINTS

- Growing bacteria in a liquid culture simplifies taking samples of the culture and determining its growth rate.
- Most bacteria grow by cell division, resulting in an exponential increase in the number of cells in a culture.
- When nutrients are plentiful, exponential growth is also balanced growth: the average amounts of the key constituents of the cells remain constant over time.
- The generation time is the time required for the number of cells to double. During balanced growth, the generation time remains constant, while its value depends on the conditions of growth and the genetic makeup of the cells.
- The generation time of a culture is most easily determined from a semilogarithmic plot of the number of cells versus the time.
- Measuring the turbidity of a liquid culture is a convenient way to follow growth.
- Liquid cultures do not represent the normal conditions of most bacteria.

Enter almost any microbiology laboratory, and you are likely to find bacteria growing suspended in liquid medium. Bacteria growing while suspended in liquid is referred to as **planktonic growth**. If the cells grow best in the presence of oxygen, they are often cultured in flasks or tubes on a platform shaker, which rotates to aerate the cultures (Fig. 3-1). There are good reasons to grow bacteria in liquid medium.

1. All the cells are in the same environment. Compare this to the growth of colonies on agar medium. Cells at the edge of the colony are in a different environment than those in the interior, so it is hard, if not impossible, to sample uniformly. However, two samples taken at the same time from a liquid culture will be identical if the culture is uniformly mixed.
2. The growth of most bacteria in liquid culture can be easily followed by measurements of optical density (turbidity).
3. Many bacteria enter a phase of exponential growth in liquid culture where the average rate of cell division during this time is constant and reproducible for the same conditions of culture.
4. Cells growing exponentially at a constant rate are in **balanced growth**: that is, to a very good approximation, all the major constituents of the cell, such as DNA, RNA, and protein, are present throughout in the same relative amounts.

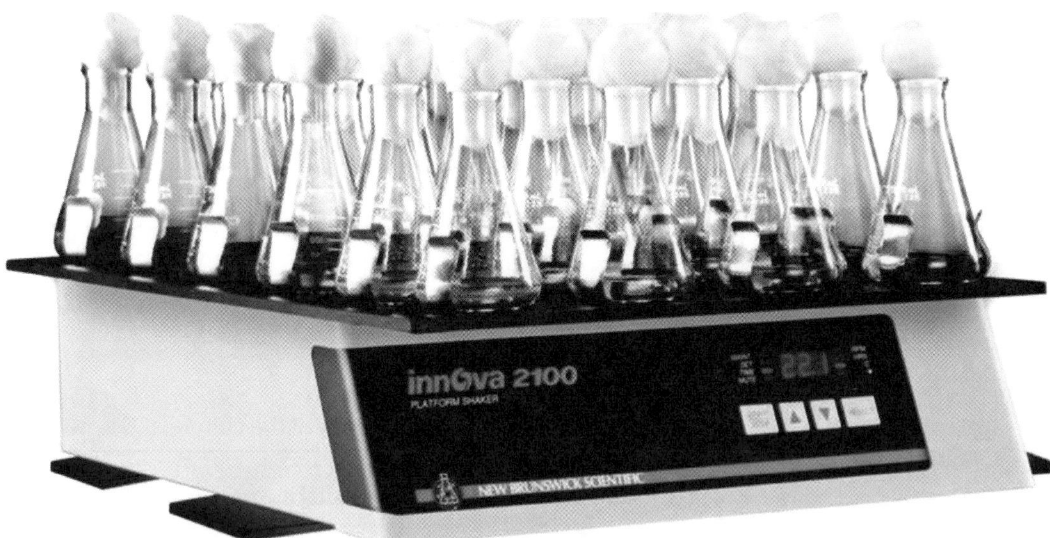

Figure 3-1 A platform shaker for growing bacteria in liquid medium. The platform rotates to improve aeration. Notice that only a small fraction of the volume of the flask is used. This is necessary to provide maximum exposure of the cells to the air. Shaker image courtesy of Eppendorf North America.

This means that cells sampled at different times during exponential growth will have the same average properties and will remain the same average size, even though the number of cells is increasing.

Bacterial growth can be defined in a number of ways, but most commonly it refers to an increase in cell number. This definition works well for cells that grow by binary fission: each cell divides and produces two daughter cells. As a consequence, the number of cells increases exponentially ($1 \rightarrow 2 \rightarrow 4 \rightarrow 8 \ldots$) (Fig. 3-2, left). Almost all the bacteria commonly encountered grow this way. However, there are exceptions: for example, one important group of bacteria, the streptomycetes, are filamentous and grow by extension and branching of cells, and our definition of growth cannot be applied to these organisms.

While the increase in cell number is one definition of growth, cell division is not all that is required. The mass of the culture must also increase. During **balanced growth**, cells have the same average size and composition while their number increases. This means that the total cytoplasmic volume of the cell must increase, not just cell number, and major components must continue to be synthesized at the same relative rates to maintain a constant amount of each in the cell.

For cells growing in liquid medium, there is usually a period of time when all the essential nutrients remain in large excess and the cells are in balanced growth. In that case, when a cell divides, the two daughter cells experience the same environment as the parental cell and have the same growth rate, measured as the time from one cell division to the next. This value is the **generation time** (Fig. 3-2, right). Of course, the time from one division to the next is not *exactly* the same but

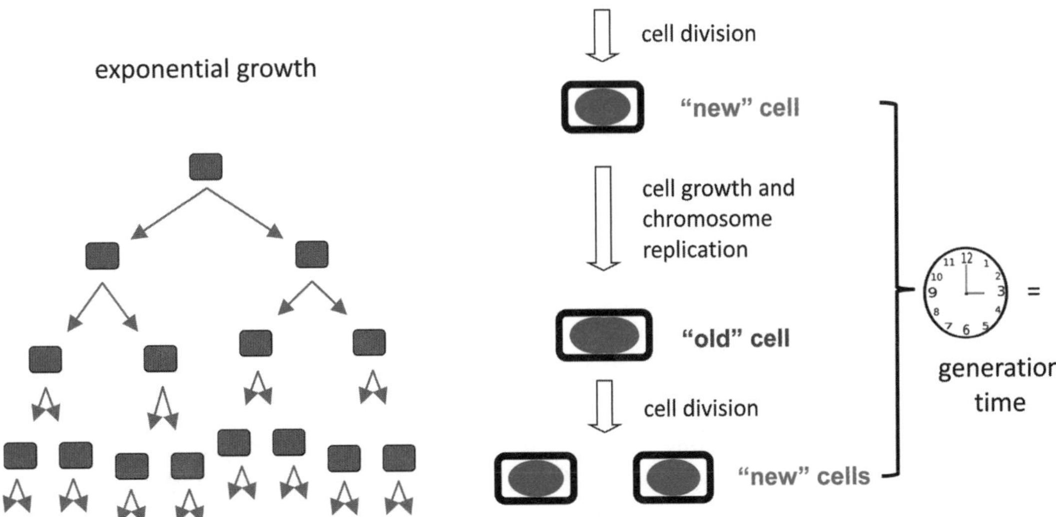

Figure 3-2 Exponential growth and the generation time.

varies slightly (Fig. 3-2, left). After many cell divisions, the result of this variation is that the cells are all dividing at different times. We can still define the (average) generation time as the time it takes for the number of cells to double. During that time, on average every cell has divided once, even though there might have been two cell divisions in some cases and none in others. Because of this definition, the generation time is sometimes called the **doubling time**.

The generation time is characteristic of a species. For example, the harmless *Mycobacterium smegmatis* has a generation time of 3 to 4 hours, whereas the generation time of *Mycobacterium tuberculosis*, which causes tuberculosis, is almost 24 hours. Within a species, bacteria with different genotypes are called **strains**. Many genes affect the rate of growth, and so different strains can have different generation times. The generation time is also affected by the culture conditions, such as temperature, type of medium, and degree of aeration. However, for any set of (unchanging) conditions, the generation time of a culture in balanced growth is constant and reproducible. The growth medium has an important impact on the generation time. A **complex medium** contains components derived from natural sources. The complex medium used in this course is tryptic soy broth (TSB), which consists of enzymatic digests of casein (a group of proteins in milk) and soybean meal. The medium mostly contains polypeptides, which are broken down to amino acids by the cell and then used for both energy generation and biosynthesis of proteins and other molecules. Many species of bacteria can grow on this or similar rich media.

A medium that contains only what is required for the growth of a particular microorganism is referred to as a **minimal medium**. Microbiologists spend a lot of effort trying to determine the composition of a minimal medium for different species of bacteria. There are two reasons for this. First, minimal medium is a **defined** medium: we know its exact composition, and every time it is made up it will be the same. The composition of complex media such as TSB will vary because no two batches of the natural components will be exactly the same. This variability could affect the characteristics of growth. Second, knowing the metabolic requirements of a species helps us to specify the conditions necessary for growth. For example, this information contributes to our understanding of what pathogens need to cause disease, where they are likely to flourish, and ways in which they might be vulnerable. A defined medium has not been determined for every species of bacteria. *Borrelia burgdorferi*, which causes Lyme disease, is in this group.

For some bacteria, a single carbon source and inorganic salts (to provide phosphorus, nitrogen, and other essential elements) are all that is required for a minimal medium. In that case, the bacteria are called **prototrophs**: they are able to make all the organic components of the cell "from scratch." Some strains of *Escherichia coli* are prototrophs. In other cases, a single carbon source is insufficient and other organic molecules must be provided. Because of several mutations, the *E. coli* strain C600 requires the amino acids threonine and leucine as well as thiamine (vitamin B_1). These compounds must be part of the minimal medium for that strain. In other cases, such as *M. tuberculosis*, the organism is naturally fastidious and a defined medium consists of many components (Dubos and Middlebrook, 1947).

The generation times of *E. coli* in complex and three different minimal media are shown in Table 3-1. It is not surprising that the cells grow fastest in complex medium, since much of what is required for their metabolism is being provided. In minimal medium, the growth rate for *E. coli* depends on the carbon source: when glucose is replaced with succinate or acetate, the generation time increases (Table 3-1). Therefore, using these media, we can show that glucose is a preferred carbon source for *E. coli*.

When cells are growing exponentially with a constant generation time, the number of cells in the culture (or their concentration) is expressed by the equation

$$N = N_0 \times 2^{(T/g)}$$

where N_0 is the starting number or concentration of cells, T is how long the cells have been growing, and g is the generation time (the time it takes for the number of cells to double).

This equation makes intuitive sense: T/g is how many times the cell number has doubled during the time T. For example, if $T = 120$ minutes and $g = 60$ minutes, then the cell number has doubled twice. If the starting number of cells $N_0 = 10^6$, then the number of cells is 2×10^6 after one doubling and 4×10^6 after two doublings, or, according to the equation

$$N = N_0 \times 2^{(120/60)} \text{ or } N = N_0 \times 2^2 \text{ or } N = N_0 \times 4$$

Since the generation time is constant and reproducible for a strain growing exponentially under defined, constant conditions, g is a useful shorthand way of indicating how fast a culture is growing. In addition, using the equation above, we can

Table 3-1 Generation times of *Escherichia coli* grown at 37 °C in different media.

Medium	Generation Time (Min)
Complex:	
Tryptone Soy Broth	30
Minimal defined:	
Inorganic Salts + Glucose	47
Inorganic Salts + Succinate	115
Inorganic Salts + Acetate	139

Table 3-2 Increase in cell number for an exponentially growing culture of *E. coli.*

T (time) min	N (cells per ml)
0	4.0×10^4
26	7.3×10^4
58	1.5×10^5
102	4.2×10^5
121	6.5×10^5
148	1.2×10^6
179	2.5×10^6

calculate the concentration of cells in the culture at any time, as long as we know the initial concentration and the generation time. We can also calculate the generation time if we know the initial and final concentration of cells and the time period of growth.

Instead of entering values into the equation, growth data are usually plotted graphically. As an example, suppose you obtain the data in Table 3-2 for a growing culture of *E. coli.* When the concentration of cells is plotted versus time, it becomes clear that because of exponential growth the number of cells not only increases with time but does so at an increasing rate (Fig. 3-3). Because of this, it can be difficult to choose a proper scale on the *y* axis. In this example, the scale is set to steps of 0.5×10^6, which is fine for most of the data but too large for the zero time point value of 4×10^4, which on this scale is very close to 0. If we used a smaller scale (for example, steps of 0.5×10^4), the *y* axis would become unmanageably large. There is a better way to plot the data in Table 3-2, by plotting *N* on a semilogarithmic graph, where the *y* axis is a logarithmic scale and the *x* axis a linear scale.

Figure 3-3 Plot of cell concentration versus time (data: Table 3-2).

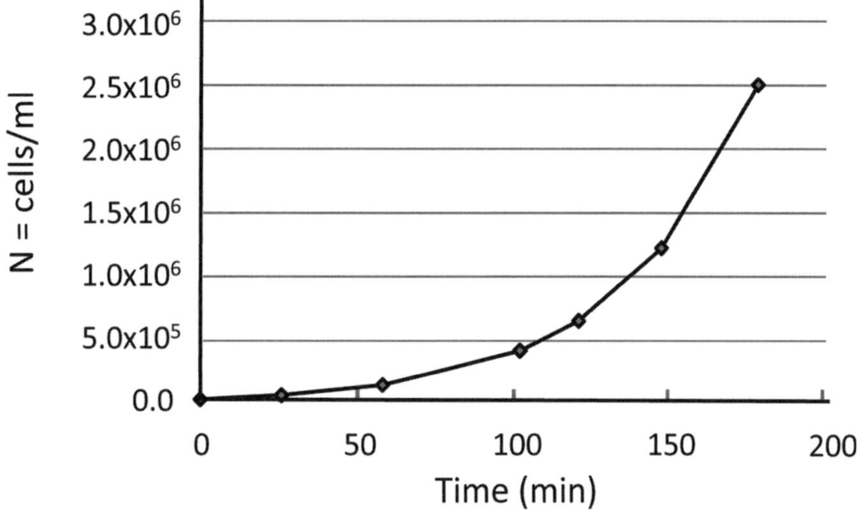

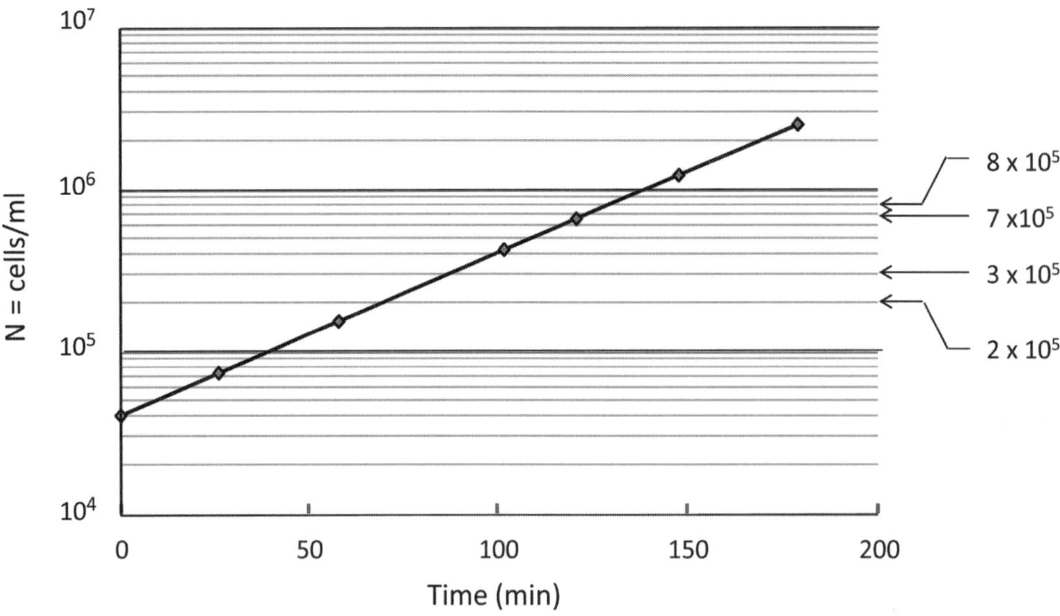

Figure 3-4 Plot of data in Table 3-2 with a log scale for the y-axis.

The advantage of the logarithmic *y* axis is that a wide range of values can be plotted on a practical scale. We can plot *N* directly by using a logarithmic scale for the *y* axis (Fig. 3-4), where the spacing on the scale is according to the logarithmic values of the numbers on the *y* axis. Consider the two pairs of *y*-axis values 2×10^5, 3×10^5 and 7×10^5, 8×10^5, marked at the right in Fig. 3-4. The difference between the values in each pair is the same, 1×10^5, so on a linear scale the distance between them would also be the same. However, the difference between the logarithms of the two numbers in each pair is not the same. For the first pair the difference is $\log(3\times10^5)—\log(2\times10^5)$, or $5.477—5.301=0.176$, and for the second pair it is $\log(8\times10^5)—\log(7\times10^5)$, or $5.903—5.845=0.058$. As a result, the distance between 2×10^5 and 3×10^5 on the *y* axis is greater than the distance between 7×10^5 and 8×10^5, in proportion to the differences in the logarithmic values. Two things to keep in mind are that (i) the repeating cycles on the scale each represent a factor of 10 and (ii) there is never a "0" on the *y* axis because log(0) is undefined. In this example, the *x* axis crosses the *y* axis at $y=1\times10^4$.

In a semilogarithmic graph of cell growth such as shown in Fig. 3-4, the greater the rate of growth, the steeper the slope of the plot. A nice feature of the plot is that the generation time can be obtained directly from the graph, without additional calculations. This is shown in Fig. 3-5. One generation is the time required for the number of cells to double. In the figure, there are 1×10^5 cells/ml at 39 minutes and 2×10^5 cells/ml at 69 minutes (indicated by the range of the brown band), so $g=69—39=30$ minutes. It does not matter where on the graph you determine *g* because it is constant throughout balanced, exponential growth. Thus, there are 3×10^5 cells/ml at 84 minutes and 6×10^5 cells/ml at 114 minutes (indicated by the range of the blue band), in agreement with $g=30$ minutes.

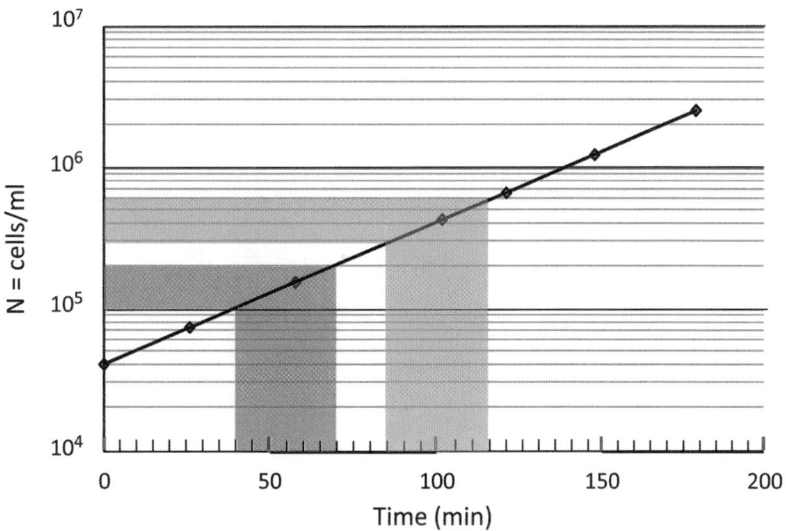

Figure 3-5 Determining the generation time from a semilogarithmic plot.

In order to generate a semilogarithmic graph like the one shown in Fig. 3-5, we need to know the number or concentration of cells when they have grown for the time period *T*. Of course, there is no problem determining *T*; all we need is a clock or timer. To determine *N*, we can count the number of cells directly under the microscope by using a special slide, called a hemocytometer, which has a well divided into sections. The slide is designed so that the volume of each section is known. By counting the bacteria in one of the sections, we can determine the concentration of cells in the culture. This is a tedious and time-consuming process. A simpler way of following the growth of a culture is to take advantage of the fact that as the number of cells increases, so does the turbidity of the culture. In fact, the turbidity, measured as optical density (OD), is proportional to the cell concentration for a range of values: that is, $\mathrm{OD} = k \times [\text{cell concentration}]$, where k is a constant number. OD can be measured using a simple spectrophotometer (Fig. 3-6). It is important to

Figure 3-6 Schematic of a spectrophotometer. Light passing through the cuvette is converted to an electric signal and amplified. The signals for the sample containing liquid culture (bacteria present) and for a blank containing the medium only (no bacteria) are compared. The decrease in signal due to light scattering by the bacteria (represented by the dashed, blue lines) is reported as OD or absorbance. Figure modified and used with permission of Prof. Friedrich Widdel, U. Bremen, Germany.

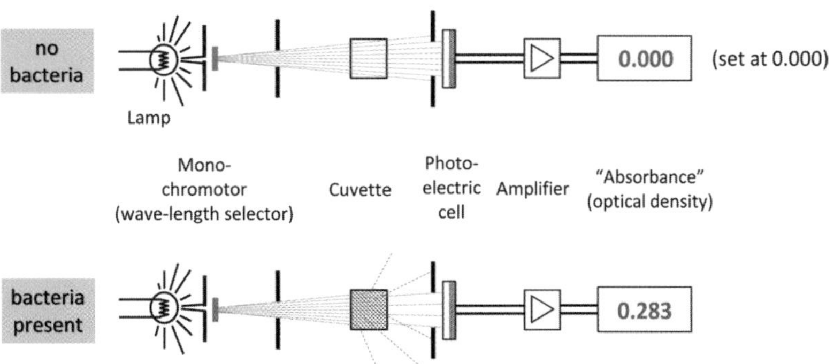

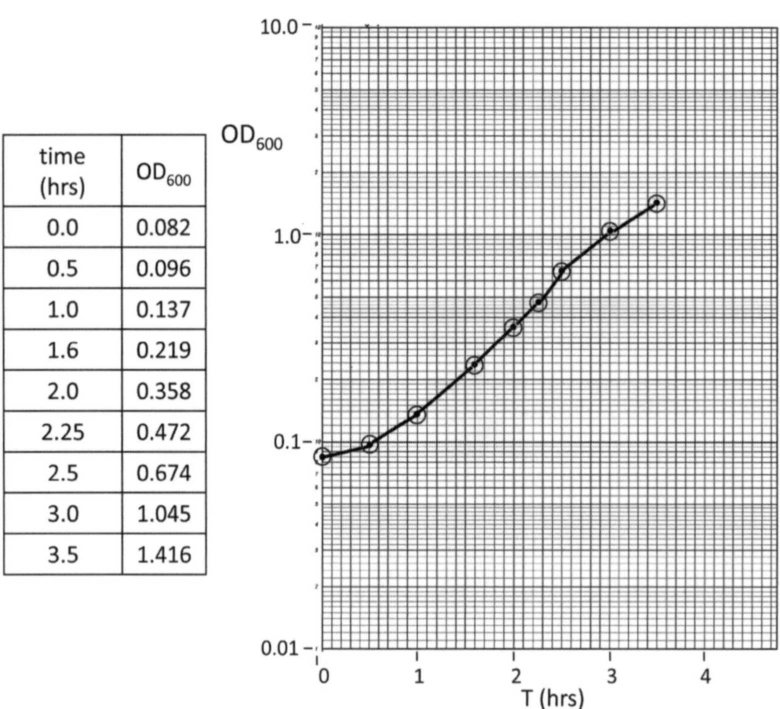

time (hrs)	OD_{600}
0.0	0.082
0.5	0.096
1.0	0.137
1.6	0.219
2.0	0.358
2.25	0.472
2.5	0.674
3.0	1.045
3.5	1.416

Figure 3-7 (left) Measurements of optical density v. time for a growing culture. (right) Data plotted on semilogarithmic graph paper. Cells are in exponential phase between approximately $OD_{600} = 0.137$ and $OD_{600} = 0.674$.

know what is being measured here. Most often, a spectrophotometer is used to measure absorbance: how much light is absorbed by a solution at a particular wavelength. The turbidity of bacterial cultures is *not* due to the absorption of light by the sample. Instead, it is due to light scattering by small particles (in this case bacterial cells), which results in light being deflected out of the pathway for detection by the spectrophotometer (Fig. 3-6). The decrease in the amount of light may be reported by the spectrophotometer as an absorbance value, but the missing light has not been absorbed; it is just not detected by the instrument. OD measurements are simple and rapid, allowing growth to be monitored while it is occurring (Fig. 3-7). The measurements are taken periodically and the time of sampling recorded. The intervals between sampling times do not all have to be the same, and because we are using a liquid culture, a 1-ml sample in a cuvette accurately represents the culture as a whole. In addition, since OD is proportional to cell number for a range of concentrations, the OD value can be plotted on semilog paper and the time it takes to go from OD value to 2×OD value is the generation time. Try the method shown in Fig. 3-5 to determine the doubling time of the organism used for Fig. 3-7. How long does it take for the organisms to go from OD = 0.2 to OD = 0.4? How about OD = 0.3 to OD = 0.6? Are the numbers similar?

The bacterial growth curve

So far, we have considered the properties of a liquid culture where the cells are growing exponentially under constant conditions. Eventually, one or more essential components of the medium are used up, exponential growth ceases, and the cells enter stationary phase (Fig. 3-8). Stationary phase is not simply a time of cell death by starvation. In *E. coli*, the stress of starvation results in the expression of new genes that enable a fraction of the cells to survive. After cessation of growth (stationary phase I in Fig. 3-8), there is a period of cell death followed by a second stationary phase. Cells in this phase are quiescent (nongrowing) and can persist for a long time. However, these cells are not a static population: during this time, mutants appear that are better able to manage the stress. Stationary phase is in fact a very dynamic period in which the cells are adapting to the harsh conditions.

Notice in Fig. 3-8 that viable (living) cells are plotted on the *y* axis. During exponential phase, practically all the cells in the culture are viable and OD measurements reflect the number of living cells. When cells enter stationary phase, a fraction of the cells die. If these cells remain intact, they will still scatter light and therefore contribute to the OD of the culture. The OD_{600} (OD at 600 nm) of the culture could remain constant while the number of living cells is decreasing.

After inoculating cells into liquid medium, there can also be a lag phase before the cells enter exponential growth (Fig. 3-8). Cells adjust to the new growth conditions during this time. A lag phase is often observed when stationary-phase (I) cells are diluted into fresh medium. This is testimony to the fact that stationary-phase cells are not simply log-phase cells that have stopped growing. Instead, they have a distinct metabolism that needs to be reset before growth can resume. In contrast, when

Figure 3-8 Generalized growth curve for *E. coli*.

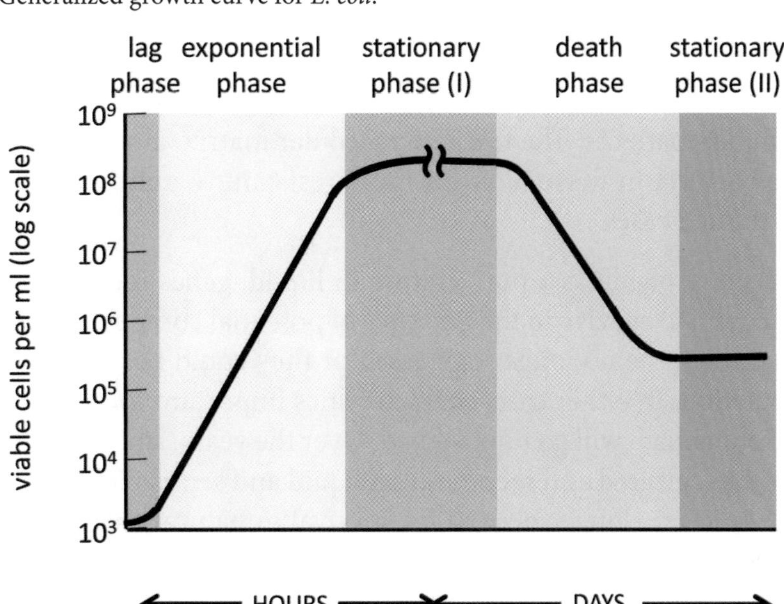

log-phase cells are diluted into fresh medium under the same conditions for growth, there is no lag phase.

There is another situation where lag phase is observed. When cells are diluted from rich, complex medium into a simple medium with a single carbon source, the biosynthetic demands on the cell are increased. As a result, new enzymes have to be synthesized before growth can resume.

Pure cultures in liquid medium and the real world of bacteria

The use of pure cultures consisting of a single species of bacteria grown in liquid medium is a mainstay of microbiology. Nevertheless, it is worth thinking about some of the limitations of this practice.

1. Liquid medium is not the usual or only growth environment for many bacteria.
2. Bacteria rarely exist as pure cultures in nature.
3. Much of the time microorganisms are in unfavorable environments and growth is limited.

For some bacteria, growing while suspended in liquid is the usual situation. The oceans contain an enormous number of bacteria growing planktonically, and together these make up the largest microbial community on the planet (Whitman et al., 1998). However, in most aqueous or semiaqueous environments, bacteria are found attached to solid surfaces by an extracellular matrix. Often this matrix hosts a community of different microorganisms, communicating and interacting in a variety of ways. These matrix-embedded communities are called **biofilms**. In unpolluted freshwater streams, the great majority of bacteria grow this way, attached to rocks, leaves, and other solid substrates. Dental plaque on the surface of teeth is another example of a biofilm. Many pathogenic bacteria form biofilms that contribute centrally to the development of disease (Challenge Six). For example, cells of *Pseudomonas aeruginosa* and other organisms form persistent biofilms in the lungs of cystic fibrosis patients. The thick extracellular matrix, along with other factors, makes the bacteria in these biofilms more resistant to antibiotics, complicating treatment in such cases.

When cells are growing as a pure culture in liquid, genes required to adhere to a solid surface and to survive in the presence of potential competitors are not needed. These genes might be no longer expressed or they could become inactivated over time by mutation. In either case, characteristics important for the natural survival of the microorganism will go unobserved. Over the years, laboratory strains of bacteria have been cultured and recultured in liquid and semisolid media. During that time, those genes no longer needed for survival in natural environments have become unnecessary baggage. The laboratory strain *E. coli* K-12 has been used for many years and has certainly lost genes required for pathogenesis. On the other hand, genes for adherence to the gut wall and entry into cells are present but not expressed. For an interesting account of this, see http://schaechter.asmblog.org/schaechter/2011/10/the-fa%C3%A7ade-of-e-coli-k-12.html.

Finally, is a prolonged period of exponential growth what bacteria normally experience in nature? In fact, most of the time bacteria are living under conditions that are far from ideal. This is why *E. coli* and many other bacteria have highly developed systems that allow them to survive for a long time under unfavorable conditions, such as those in stationary phase II (Fig. 3-8). Even more impressive are the spores of bacilli, which might be able to survive for millions of years (Challenge One, Cano and Borucki, 1995). The presence of these complex survival systems reflects the fact that bacteria may have to endure long periods of deprivation, yet microbiologists until recently have focused on the properties of exponential-phase cells.

Lab One

1. ***Construct a graph of bacterial growth by measuring optical density.***
2. ***Determine the concentration of viable cells by serial dilution and plating.***
3. ***Calculate the relationship between optical density and viable counts for exponential-phase cells.***
4. *Determine the minimal inhibitory concentration for different antibiotics by 2-fold dilutions in broth medium.*
5. *Interpret the results to identify the antibiotics that might be useful for treatment.*

Learning outcomes

After this lab, students will be able to:

- **a.** Use a spectrophotometer.
- **b.** Construct a growth curve on semilogarithmic paper.
- **c.** Explain the purpose of a dilution series.
- **d.** Do a series of 10-fold dilutions.
- **e.** Use colony counts to calculate the concentration of viable cells in a liquid culture.
- **f.** Determine the doubling time of an organism from a growth curve.
- **g.** Describe the relationship between optical density and the number of viable cells in a culture.

I. Construct a growth curve and calculate the generation time

At the beginning of the lab, the instructor will add an overnight culture of the *Pseudomonas* strain into one or more flasks containing fresh broth so that the cells are diluted 100-fold (i.e., 0.1 ml culture will be added for every 9.9 ml broth). The diluted cultures will then be transferred to a water bath-shaker set to 30°C. The instructor will immediately take a sample from one of the flasks and demonstrate how OD is determined by using the spectrophotometer in the laboratory. Notice that a cuvette containing medium only will be used first to blank the spectrophotometer. This is necessary to ensure that only light scattering by the cells is being measured, not absorbance by the liquid broth as well.

PROCEDURE

Recording the OD of a growing culture

1. One culture will be available for every group of four to six students (two to three pairs/group). Each group should have a copy of Table 3-3. Record the culture you are using in the upper left corner of the table.
2. Each pair should obtain a 1-ml disposable cuvette.
3. The outline for sampling the growing culture is shown in Fig. 3-9. Your group should plan on taking samples at approximately uniform intervals. For example, if 150 minutes are available to incubate the cells and there are three pairs per group, sample at 0 minutes and every 30 minutes thereafter. These intervals do not have to be exact, but the time each sample is taken should be accurately recorded. As shown in Fig. 3-9, the pairs in the group should sample in the same order so that there is maximum time for each pair to do the serial dilutions (below).
4. For each sample, remove 1 ml of the culture and add it to a cuvette. Record the time in Table 3-3. First, set the spectrophotometer reading to 0 absorbance by using the blank cuvette (medium only), which will be provided by the instructor and placed next to the instrument. Then insert your sample cuvette and record the OD_{600} in the column labeled "(A) OD_{600}" in the table.
5. Transfer 0.1 ml of the sample from the cuvette to a tube containing 0.9 ml sterile water for serial dilution (below).
6. Pour the remaining sample back into the flask and continue the incubation.

   Normally, samples are discarded after determining the OD_{600} to maintain the sterility of the culture. Sterility is not essential here because the short time of growth and large inoculum means that the effect of any contaminant will be negligible.
7. Rinse the cuvette several times with distilled water over the designated container, shake to remove any large drops of liquid, and store the cuvette inverted on a lab wiper for drying. Be sure there are no large drops in the cuvette before you take your next sample.
8. As you proceed, use the data in Table 3-3 to plot the OD versus time, with OD_{600} on the *y* axis. You can use the semilog graph paper provided at the end of this section. Connect the points with a line. When the cells are in exponential phase (approximately an OD_{600} between 0.1 and 0.8), the plot will be close to a straight line. Figure 3-7 (right) is a sample plot.

Table 3-3 Data for calculation of cell concentration at $OD_{600} = 1$.

Culture used: ________________

			Colonies per plate:				(B)	
Pair	Time (min)	(A) OD_{600}	10^{-5}	10^{-6}	10^{-7}	10^{-8}	Viable cells per ml	(B) / (A) Viable cells per ml at $OD_{600} = 1.0$
1	0							
2								
3								
1								
2								
3								
1								
2								
3								

When all the data are entered in Table 3-3, you will be able to estimate the generation time of the exponential-phase cells from the graph, by selecting values for OD_{600} and $2 \times OD_{600}$ on the y axis and their corresponding times on the x axis. Do this for two different pairs of time points. If the values for the generation time are nearly the same for both pairs, then the cells are in exponential phase under constant conditions (Fig. 3-5).

II. Determine viable cell counts during exponential growth

Using the $T = 0$ sample, your instructor will demonstrate serial dilutions and the spreading technique. Using OD to monitor the growth of exponential-phase cells is simple and convenient. However, we are usually interested in the concentration of viable cells, not OD, so we need to establish the relationship between the two. The number of living bacteria is determined by plating the culture onto agar medium and then counting the number of resulting colonies. Since each isolated colony originates from a single cell, the number of colonies represents the number of cells initially deposited on the plate. The problem is that bacterial cultures usually contain large numbers of organisms (10^6 cells per ml or more). Normally, 0.1 ml of liquid is spread on a plate of agar medium. If we sampled a culture containing 10^6 cells per ml, we would be depositing $0.1 \times 10^6 = 10^5$ cells on the plate, which would form a semiconfluent lawn rather than individual colonies (which would be too

Figure 3-9 Measuring cell growth and serial dilutions: outline of work flow.

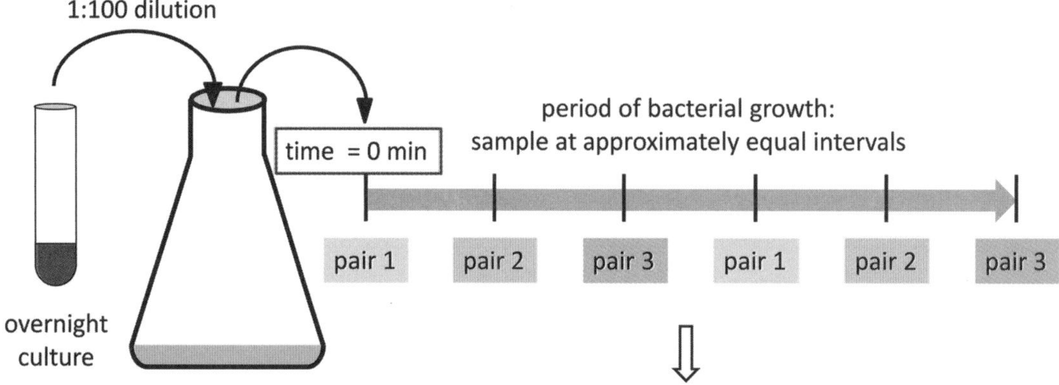

Remove 1 ml from the culture and add to cuvette.
Record the time of sampling.

Measure the OD with the spectrophotometer

Remove 0.1 ml from cuvette and add to
0.9 ml sterile water for serial dilutions

many to count in any case). Instead, the culture is diluted in a series of steps and samples of several different dilutions are then spread on agar medium. If there is a countable number of colonies on one of the plates, we can then calculate the concentration of cells in the culture if we know how we did the dilutions. Calculations are easiest if cells are diluted by a factor of 10 at each step.

To see how serial dilutions work, suppose a culture contains 4×10^6 cells per ml culture (Fig. 3-10). In that case, a 0.1-ml sample of the culture contains $0.1 \times 4 \times 10^6 = 4 \times 10^5$ cells. When this sample is added to 0.9 ml sterile liquid, the 4×10^5 cells in the 0.1-ml sample are now in 0.1 ml + 0.9 ml = 1.0 ml. The cell concentration becomes $(4 \times 10^5)/(0.9 \text{ ml} + 0.1 \text{ ml}) = 4 \times 10^5$ cells per ml. The concentration decreases by a factor of 10, a 10-fold dilution. Calculation of the dilution is easy when using 10-fold dilutions: we just need to decrease the exponent, in this case from 6 to 5. Note that a 10-fold dilution here means 1 volume of cells is added to 9 volumes of sterile liquid. In microbiology, this is the customary definition of a 10-fold or "1:10" dilution.

By repeating the procedure successively (Fig. 3-10), a set of dilutions is created where the concentration of cells in each tube decreases in 10-fold steps. This is referred to as making a serial dilution. For each tube, the overall dilution of the cells is the product of all the dilutions. For example, if the cells have been diluted twice (two dilution tubes), then the overall dilution is $1/10 \times 1/10 = 1/100$ or 10^{-2} of the original culture, for a final concentration of 4×10^4 cells per ml. If the sample is further diluted in 10-fold steps, then the overall dilutions are progressively 10^{-3}, 10^{-4}, and so on.

Figure 3-10 Result of serially diluting a culture with a known concentration of cells.

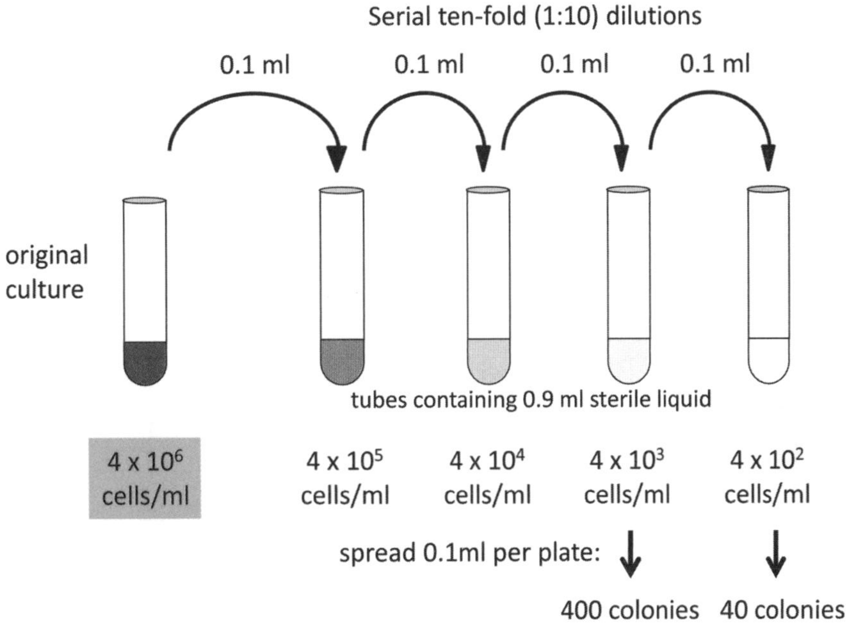

When the culture is sufficiently dilute, a sample can be spread on agar medium and the number of colonies counted. Only 0.1 ml of a dilution is used for spreading. Therefore, if the concentration of cells is 4×10^3 per ml in one of the dilution tubes (Fig. 3-10), $0.1\times4{,}000=400$ cells will be deposited on the plate and we would expect 400 colonies (the actual number will vary somewhat).

Using serial dilutions, we can determine the concentration of viable (colony-forming) cells in a culture. First, we create a set of serial dilutions as before. Since we do not know which dilution will be best, 0.1-ml samples of a number of different dilutions are each spread on agar medium and incubated until colonies appear. As an example, an overnight culture of *E. coli* was serially diluted and 0.1 ml of dilutions 10^{-5}, 10^{-6}, 10^{-7}, and 10^{-8} then spread on agar medium (Fig. 3-11). The next day, the number of colonies on each plate is counted (Fig. 3-12). The highest number of countable colonies gives the most accurate result, so we select the plate containing 79 colonies. Since 0.1 ml of the 10^{-7} dilution was spread on this plate, the concentration of cells in the 10^{-7} dilution must be $10\times79=790$ cells per ml. There are then $10\times790=7{,}900=7.9\times10^3$ cells per ml in the 10^{-6} dilution, 7.9×10^4 cells per ml in the 10^{-5} dilution, and so on. It is easy to see why there are so many colonies on the 10^{-5} plate, because $0.1\times7.9\times10^4=7{,}900$ cells were deposited on the plate. Calculating up in 10-fold steps from the cell concentration in the 10^{-7} dilution, we can arrive at the concentration of viable cells in the overnight culture. Alternatively, because the cumulative dilution of the culture was 10^{-7} prior to spreading the cells, the concentration of cells is simply $10\times79\times10^7=7.9\times10^9$ cells per ml. Of course, if the dilution series were repeated, or colonies from a different plate were used, a slightly different number for the cell concentration would be obtained, due to the small variations in the amount sampled at each step. In practice, plates having between 30 and 300 colonies are usually selected for colony counting.

Figure 3-11 Ten-fold dilution and plating an overnight culture of *E. coli.*

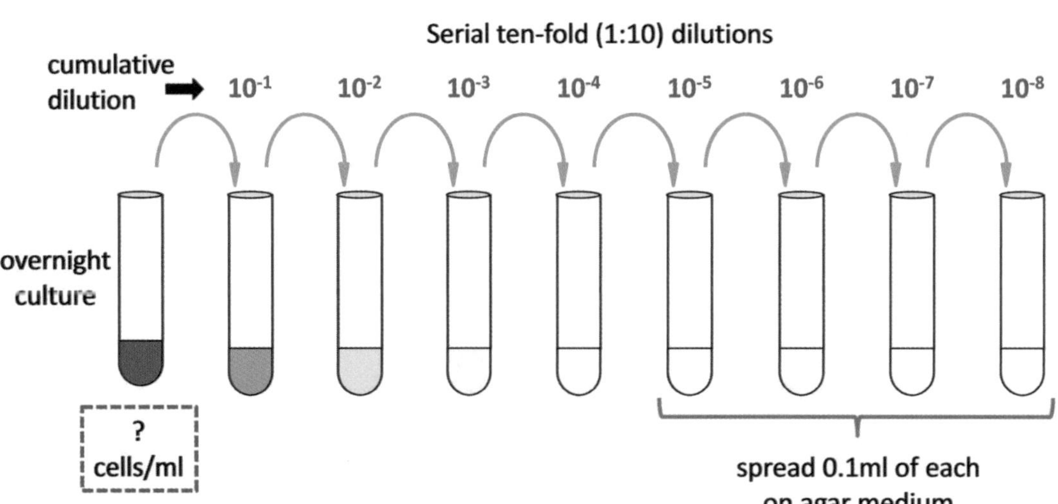

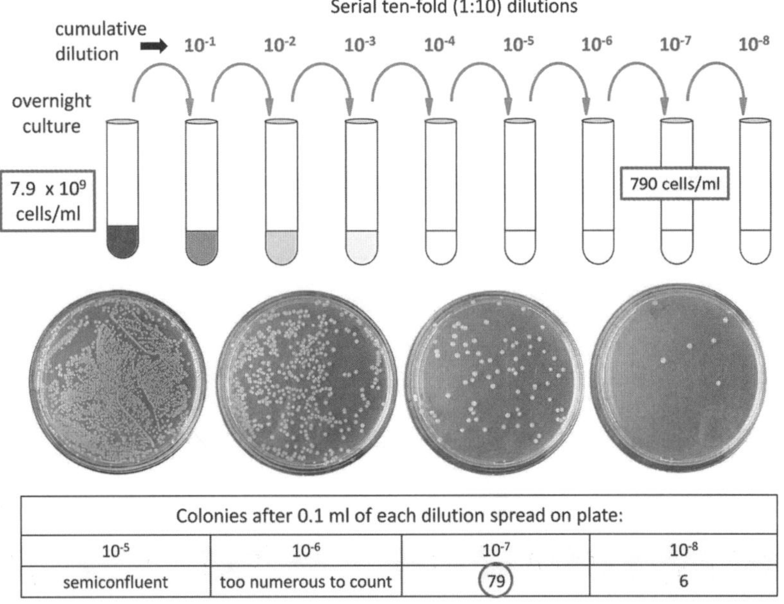

Colonies after 0.1 ml of each dilution spread on plate:			
10^{-5}	10^{-6}	10^{-7}	10^{-8}
semiconfluent	too numerous to count	(79)	6

Figure 3-12 Results of serial dilution.

PROCEDURE

Serially diluting your samples

1. You will do a serial dilution for each of your samples from the growth curve. Notice that you have already diluted the culture 1:10 by adding 0.1 ml of the culture to a 0.9-ml sterile water blank.
2. The blanks may be provided, or your instructor may ask you to make the blanks yourself using sterile tubes and water or medium. Dilute the cells in 10-fold steps so that you have the following:

   Sample $OD_{600} < 0.20$: tubes containing 10^{-5}, 10^{-6}, and 10^{-7} dilutions

   Sample $0.20 < OD_{600} < 1.0$: tubes containing 10^{-6}, 10^{-7}, and 10^{-8} dilutions

   Notice that when the OD_{600} is greater than 0.2, higher dilutions are used. This is because the OD_{600} is proportional to cell concentration: as OD increases, so does the concentration of cells, and therefore higher dilutions are needed to obtain a countable number of colonies.
3. Before beginning, determine the dilutions you will be doing according to the OD_{600} of the sample. Sketch the blank tubes you will be using and the dilution at each step, as in Fig. 3-12.

Good sterile technique for doing serial dilutions:

1. Always use a new pipette tip for every step in the dilution.
2. Make sure you are sampling the correct volume. Normally, this will be 100 µl (0.1 ml). Familiarize yourself with what this volume looks like in a pipette tip, and then visually confirm that you are pipetting the correct amounts while doing the dilution series.
3. Make sure you empty the contents of the tip completely into the blank tube. If the pipette tip is loose or if the part of the tip that fits on the pipette is out-of-round, the correct volume will not be drawn up and/or delivered, even if the pipette has been correctly set.

Procedure continues on next page

Serially diluting your samples (*continued*)

4. Gently mix the contents of the tube after you have added the cells. Vigorous vortexing is not required, but make sure that the side of the tube is washed by shaking the tube a few times.
5. Keep track of your diluting. A customary way to do this is to line up the dilution tubes in the front of your rack. When you have sampled from a tube, move the tube in some regular way (for example, back one row) to indicate that it has been used. It is also a good idea to label each tube with the intended dilution.
6. Make sure the tubes are capped except when you are adding or removing liquid. Don't place the cap on the bench surface; instead, pull off the cap and hold it with your small fingers, as described in Challenge One.

PROCEDURE

Spreading cells on agar medium

1. Remember, you will be spreading the following:

   Sample $OD_{600} < 0.20$: 10^{-5}, 10^{-6}, and 10^{-7} dilutions

   Sample $0.20 < OD_{600} < 1.0$: 10^{-6}, 10^{-7}, and 10^{-8} dilutions
2. Deposit 0.1 ml of each dilution on a separate tryptic soy agar (TSA) plate. Do not use wet plates: examine the plate to make sure there is no liquid on the surface of the agar medium.
3. For $OD_{600} < 0.2$, spread 0.1 ml from the 10^{-5}, 10^{-6}, and 10^{-7} dilutions on TSA plates. For $OD_{600} > 0.20$, spread 0.1 ml from the 10^{-6}, 10^{-7}, and 10^{-8} dilutions. Incubate the plates at 30°C.

There are two commonly used ways to spread cells:

L-rod technique

1. You will be given a glass L-rod for spreading the cells and a small glass container with ethanol.
2. Holding one end of the L-rod, dip the other into the ethanol, tap on the top of the container to remove excess liquid, and then pass the L-rod through a Bunsen burner.

   The alcohol will ignite and dust particles and other matter will be burned off, leaving a sterile surface. Do not hold the L-rod in the flame so that it becomes very hot. The L-rod must be near room temperature before spreading or it will kill the cells. The L-rod will become cool in about 10 seconds if it has been flamed properly.

SAFETY Never leave flame unattended. Make sure flame is extinguished and gas supply is off before leaving the lab. Ethanol is highly flammable and burns with a nearly colorless flame that can go unnoticed. Before placing the L-rod back into the ethanol, make sure the flame on the surface of the L-rod is completely extinguished. Never hold an L-rod with burning ethanol over the ethanol container.

3. Spread the cells on the plate using a circular motion. One way of doing this is to spread back and forth in one direction while rotating the plate. Continue until all the liquid sample is absorbed by

the plate. Make sure to spread the liquid evenly over as much of the agar surface as you can. Remember, you want to obtain well-isolated colonies.

4. Again sterilize the spreader by dipping in alcohol and igniting. After the L-rod has cooled, you can place it on the lab bench.

Although the L-rod technique is widely used, it is hazardous because the ethanol in the container can accidentally ignite while you are flaming the L-rod. Single-use, presterilized, disposable L-rods have become popular because they eliminate the need for ethanol.

Sterile beads technique

1. Add 5 or 6 sterile glass beads to the plate containing the sample.
2. Shake the plate back and forth several times, rotate the plate about one-sixth of a turn, and shake again. Avoid shaking the plate up and down as culture will splash up into the lid. Continue until the beads have uniformly contacted the entire agar surface and there is no surface moisture.
3. Pour the beads into the designated container in the lab. The beads will later be collected, washed, and sterilized.

This technique does not require flaming with ethanol, and after some practice, plates containing beads can be stacked so that more than one plate can be spread at the same time. A disadvantage is that the beads must be collected, cleaned, and sterilized before they can be reused.

During the week

1. After 18 to 24 hours, when the isolated colonies are easily visible, count the colonies on each plate where possible (countable plates have between 30 and 300 colonies). Do the dilutions look correct (approximately a 10-fold decrease in the number of colonies for each dilution step)?
2. Enter the colony count for each time point into the correct dilution column in Table 3-3. Remember, the plate used for counting should have at least 30 colonies.
3. Calculate the concentration of viable cells at each time point. Enter the result in the column "(B) Viable cells per ml."
4. The number of viable cells per milliliter (column B in Table 3-3) divided by its corresponding OD_{600} (column A in Table 3-3) is the concentration of cells per milliliter when $OD_{600} = 1$. Calculate this for as many different time points as you can and enter the values in the column "(B)/(A) Viable cells per ml at OD_{600}=1" in Table 3-3. During balanced growth, (B)/(A) will be roughly the same. You can select the value you will use as an average or from a time point with the greatest number of countable colonies.

You are now able to calculate the OD at a particular cell density. Suppose you find that $OD_{600} = 1$ corresponds to 2×10^9 cells/ml. You want a concentration of 5×10^8 cells/ml. What will be the OD_{600} of a culture with this concentration of cells? Since OD is proportional to cell concentration:

$$\frac{OD_1}{OD_2}=\frac{C_1}{C_2} \text{ so } \frac{OD_{600}}{1}=\frac{5\times10^8}{2\times10^9}$$

The desired $OD_{600} = 0.25$.

QUESTIONS

Questions are designated B1 to B6 according to the six levels of Bloom's taxonomy.

1. When *Serratia marcescens* is grown in liquid medium at 25°C, it produces the red pigment prodigiosin. How might this complicate measurements of OD with the spectrophotometer? (B2)
2. In semilogarithmic plots like the one in Fig. 3-4, the faster the cells are growing, the steeper the slope of the line. Explain why. (B2)
3. Suppose that when cells enter stationary phase, all the dead cells remain intact. Sketch a plot of OD_{600} in Fig. 3-8. (B3)
4. If it takes 48 minutes for cells in exponential phase to increase from 3×10^6 cells/ml to 5.5×10^6 cells/ml, how long would it take for these cells to increase from 3×10^4 cells/ml to 5.5×10^4 cells/ml? (B3)
5. Plot the data in Table Q3-1 on semilog graph paper. Estimate the generation time from the resulting plot. (B3)

Table Q3-1 Growth of *E. coli* during exponential phase in complex medium

Time (hr):	Cells per ml:
0	1.2×10^4
1.0	3.7×10^4
1.5	6.5×10^4
2.0	1.1×10^5

6. You are asked to prepare for the class a mixture of *E. coli* and *Staphylococcus epidermidis* so that there are approximately equal numbers of cells each at a concentration of about 1.5×10^8 cells per ml for each and in a final volume of 10 ml. The two strains were grown separately overnight, and serial dilutions were then plated to determine the numbers of viable cells. The results are in Table Q3-2. Calculate the approximate volume of each overnight culture and the volume of fresh medium that would be required for the class mixture. (B3)

Table Q3-2 Plating results: serial dilutions of overnight cultures of *E. coli* and *Staphylococcus epidermidis*

	No. of colonies after plating 0.1 ml of indicated dilution:	
Dilution:	*E. coli*	*S. epidermidis*
10^{-5}	(too numerous to count)	459
10^{-6}	153	48
10^{-7}	14	5
10^{-8}	1	0

7. In serial dilutions, a "1:10 dilution" means adding 1 volume of the liquid containing the cells to 9 volumes of the liquid used for dilution. In chemistry, a "1:10 dilution" often means adding 1 part of a solution to 10 parts of the dilution liquid. Why is the definition used by microbiologists more convenient than the one used by chemists when doing serial dilutions? (B3)

8. "The number of viable cells per milliliter divided by its corresponding OD_{600} is the concentration of cells per milliliter when $OD_{600} = 1$." Show that this statement is true using the equation $OD_1/OD_2 = C_1/C_2$. (B3)

9. You need cells at a density of 4×10^7 where $0.8\ OD_{600} = 5 \times 10^8$ cells/ml. What will be the OD_{600} reading when the cells reach the desired concentration? (B3)

10. You plan on starting a culture, then going to lunch for an hour. If the generation time is 20 minutes, at what OD_{600} should you head for lunch so that when you return the culture is at the desired OD in question 9? Assume the cells are in exponential phase throughout. (B5)

11. Suppose during serial dilution you use a plate with a lot of moisture on the surface of the agar medium. You spread the cells and then place the still-wet plate in the incubator. What is the likely result? (B5)

12. A student sets up serial dilutions using a new pipette tip for each step. He deposits 0.1 ml of each dilution on a plate in the order 10^{-8}, 10^{-7}, and 10^{-6}. However, he forgets to change pipette tips during this step. Would a serious error be introduced into his results? Would the error be the same if the dilutions were done in the opposite order? (B5)

13. *B. burgdorferi*, the agent causing Lyme disease, is able to survive for a long time in our bloodstream. Blood is a complex medium, so you want to determine which of its components are required for survival of the bacteria. How does the lack of a defined medium complicate this investigation? What experiments would you do if a defined medium were available? (B6)

Lab Two

BACKGROUND

Assaying for antibiotic sensitivity

KEY POINTS

- An antibiotic kills bacteria or inhibits their growth by targeting a specific and important cellular process.
- Resistance to antibiotics can be innate or acquired. Acquired resistance can be due to mutation of genes already present in the cell or to entirely new genes obtained by horizontal gene transfer.
- The minimal inhibitory concentration (MIC) of an antibiotic is the lowest concentration inhibiting growth of a target microorganism. Dilution assays and Etests are two methods used to determine MIC.

Antibiotics have been a major tool in the treatment of disease. These compounds are not simply agents that are generally toxic to the cell. Instead, antibiotics have highly specific targets: different cellular processes essential for growth and survival. For example, penicillin and its many derivatives inhibit synthesis of the cell wall. Other antibiotics, such as nalidixic acid, target DNA replication, while rifampicin is an example of an antibiotic that interferes with transcription. Many antibiotics bind to the ribosome and affect translation. These include streptomycin, kanamycin, and chloramphenicol.

Although antibiotics are powerful, they cannot be used indiscriminately. There are several steps in selecting an antibiotic for the treatment of patients infected with a new strain. The first is to test the sensitivity of the microorganism to a panel of different antibiotics. Resistance to an antibiotic is frequently encountered and can be either innate or acquired. Innate resistance refers to a normal property of the microorganism that renders the antibiotic ineffective. For example, proteobacteria are naturally resistant to vancomycin because the drug is not able to penetrate the outer membrane. The important pathogen *P. aeruginosa* is notoriously resistant to many antibiotics, with sensitivity to some of these appearing only at very high concentrations of the drug. This is due to a combination of at least two factors. The outer membrane of *P. aeruginosa* is particularly impermeable to antibiotics, compared to *E. coli* and many other proteobacteria. In addition, *Pseudomonas* has efflux pumps that transport a variety of compounds, including antibiotics, out of the cell. As a result of these characteristics, the intracellular concentration of an antibiotic required for effectiveness might not be easily achievable (Livermore, 2002).

Bacteria can also acquire resistance to antibiotics. This is sometimes due to a mutation that alters the target of the drug. Trimethoprim is an antibiotic commonly used to treat urinary tract infections due to *E. coli*. The drug binds to and inactivates the enzyme dihydrofolate reductase, which is required for several important metabolic pathways, notably the synthesis of thymidylic acid (TMP), a precursor of DNA. Mutations resulting in resistance to trimethoprim have appeared in *E. coli* and other species. These mutations alter the enzyme so that it no longer binds trimethoprim but is still active for biosynthesis. In other cases, acquired resistance is the result of genes introduced by plasmids and other mobile elements. These genes often encode an enzyme that directly inactivates the antibiotic. Plasmids may contain a number of genes with each conferring resistance to a different class of antibiotics. These plasmids are a major problem because when they are acquired, the cell becomes multidrug resistant in a single step.

A common and inexpensive way to assay the response of a bacterial strain to different antibiotics is by the **disk diffusion (Kirby-Bauer) assay** (Fig. 3-13). Bacteria are swabbed onto agar medium and filter disks impregnated with antibiotic then placed on the surface. The antibiotic diffuses from the disk into the surrounding medium during incubation of the culture. If the bacterial strain is sensitive to the antibiotic, a zone of inhibited growth will appear around the filter disk. The size of the zone will depend on the level of sensitivity of the bacteria and the amount of antibiotic initially on the filter.

The simple disk diffusion assay is useful for the initial, rapid testing of sensitivity to a wide range of antibiotics. Clinically, the test is performed under highly standardized

Figure 3-13 Antibiotic susceptibility testing by the disc diffusion (Kirby-Bauer) assay. If a bacterial strain is sensitive to the antibiotic on the disc, a zone of growth inhibition appears around the disc. Modified from https://clinicalgate.com/laboratory-methods-and-strategies-for-antimicrobial-susceptibility-testing-2/. Originally from: Tille (2014).

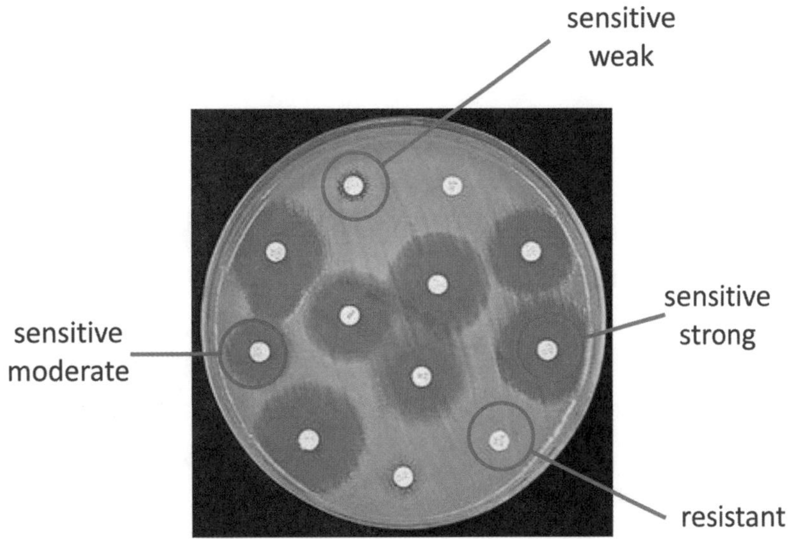

conditions that ensure reproducibility and allow comparison of the results from different labs. The test does have a shortcoming, however. Before considering an antibiotic for clinical use, it is important to know the **minimal inhibitory concentration (MIC)**. The MIC is the lowest concentration of the antibiotic that inhibits growth. MICs are not easily determined from disk diffusion assays because the concentration of antibiotic defining the zone of inhibition is unknown.

Another way of determining the MIC is to assay for growth in medium containing different, known concentrations of antibiotic (Fig. 3-14). Typically, the concentration is decreased in 2-fold steps, for example, 128, 64, 32, 16, 8, and 4 μg/ml. After incubation of the cultures, the MIC is defined as the lowest concentration of antibiotic where there is no visible cell growth. As with the disk assay, this test is done under standardized conditions. Culture tubes are often used, but the test can be miniaturized by using 96-well microtiter plates (Fig. 3-14). Instruments that determine MIC by a partially or fully automated process, requiring very small amounts of culture, are also commercially available but are expensive.

Figure 3-14 Assaying for antibiotic sensitivity with a two-fold dilution series. (top) Tube assay. Green = no growth, tan = some growth (visible turbidity) brown = growth (very turbid). (bottom) Assay using multiwell plates. Abbreviations for antibiotics are: Cm = chloramphenicol, Nl = nalidixic acid, Ap = ampicillin, Gn = gentamycin, Tp = trimethoprim. Well plate template from: Edita Aksamitiene **http://www.cellsignet.com/media/templ.html**.

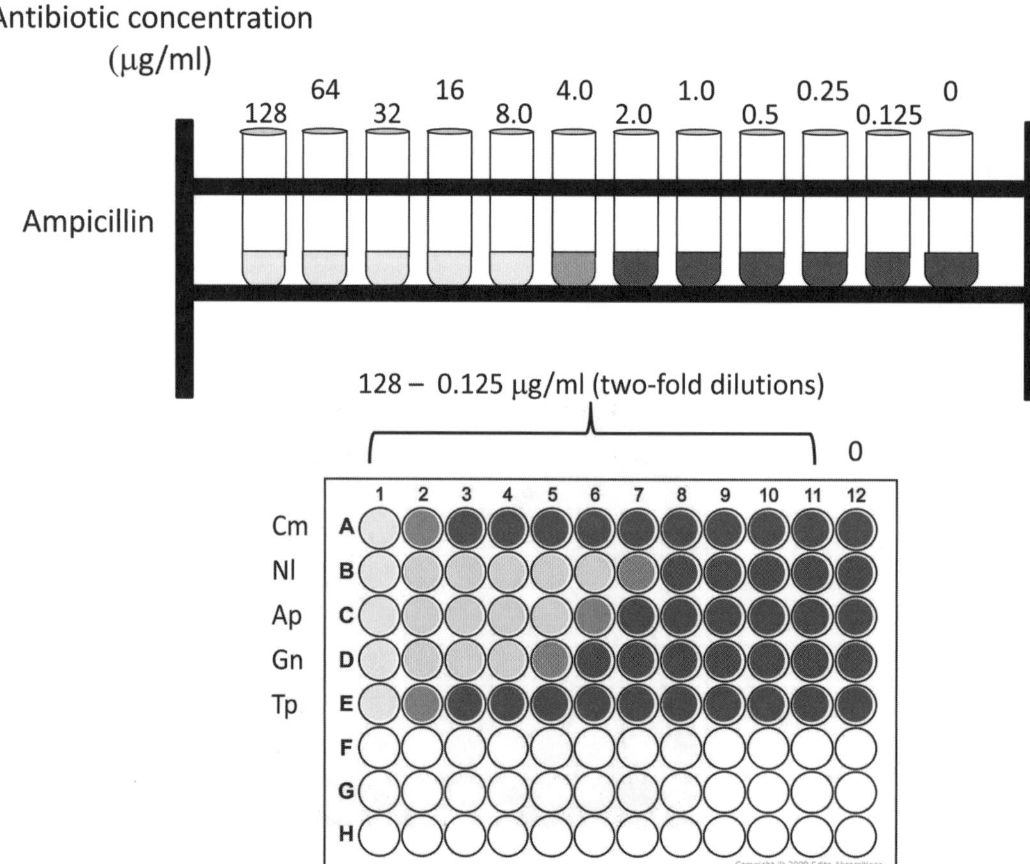

The **dilution assay** has been used for many years and is a trusted, reliable procedure for determining the MIC. However, in the absence of an automated system, it is labor-intensive and prone to human error. The **Etest** is a way of measuring MICs that retains the simplicity of the agar diffusion assay while providing a quantitative result. In this test (Fig. 3-15), a plastic strip containing a concentration gradient of antibiotic is placed on a freshly spread lawn of cells. The concentration of antibiotic at each point along the strip is indicated by a printed scale. The antibiotic diffuses from the strip into the medium, but the highest concentration of antibiotic is at the strip and indicated by the scale. If the cells are sensitive to the antibiotic, there will be a zone of inhibited growth. However, the lower the concentration, the smaller the zone of inhibition, and at some point the zone will disappear. The concentration of antibiotic at this point, which can be read directly from the strip, is taken as the MIC. In general, there is good agreement between the MICs determined by the dilution assay and the Etest, although some exceptions have been reported.

The MIC sets an important starting point in considering whether an antibiotic will be useful for treatment, but other information is needed before a decision on

Figure 3-15 The E-test® method for determining MICs. The scale is micrograms (µg) per ml of antibiotic at that location on the strip. The MIC is approximately 0.75 µg per ml. Credit: Charles University Prague http://mikrobiologie.lf3.cuni.cz/bak/uceb/obsah/disktest/etest.htm. E-test® is a registered trademark of BioMérieux (France).

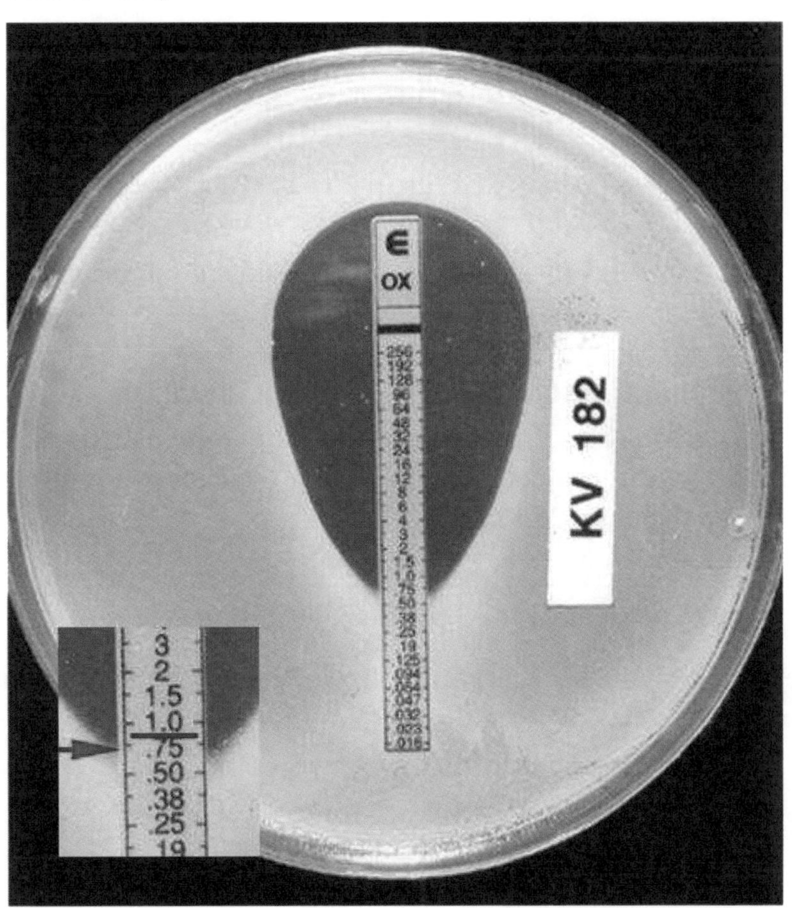

therapeutic use can be made. A very high MIC will rule out some antibiotics for consideration if these levels cannot be realistically achieved in the patient or are toxic. Even if the drug appears suitable according to its MIC and low toxicity, the minimal concentration for a favorable outcome during treatment, the "clinical breakpoint," depends on many factors, including the rate that the antibiotic is cleared from the system and the location of the infection. If the antibiotic is rapidly removed or inactivated, the necessary concentration of antibiotic might be present for too short a time to be effective. The concentration of antibiotic over time is generally monitored by examining the levels in blood serum. However, there can be much lower concentrations at some potential sites of infection, such as the cerebrospinal fluid. In summary, regimens for treatment with an antibiotic are based not only on the MIC but on pharmacological studies and clinical experience.

Lab Two

1. *Construct a graph of bacterial growth by measuring optical density.*
2. *Determine the concentration of viable cells by serial dilution and plating.*
3. *Calculate the relationship between optical density and viable counts for exponential-phase cells.*
4. ***Determine the minimal inhibitory concentrations of different antibiotics by 2-fold dilutions in broth medium.***
5. *Interpret the results to identify the antibiotics that might be useful for treatment.*

Learning outcomes

After this lab, students will be able to:

- **a.** Calculate how to prepare a culture with a specific concentration of antibiotic and cells.
- **b.** Determine the MIC of an organism using 2-fold dilutions.

I. Determine the MICs of different antibiotics for the *Pseudomonas* isolate

The class will determine as a group the MICs of the antibiotics ampicillin, kanamycin, nalidixic acid, and chloramphenicol for the *Pseudomonas* strain isolated from the patient with Crohn's disease.

The conditions for determining the MIC of each antibiotic are summarized below and are similar to the standardized protocol for this procedure:

1. The range of antibiotic concentrations for each will be 16 to 250 μg/ml, with a 2-fold decrease in concentration within this range.
2. The culture volume will be 1 ml.
3. The initial cell concentration in each tube will be 10^5 cells/ml.

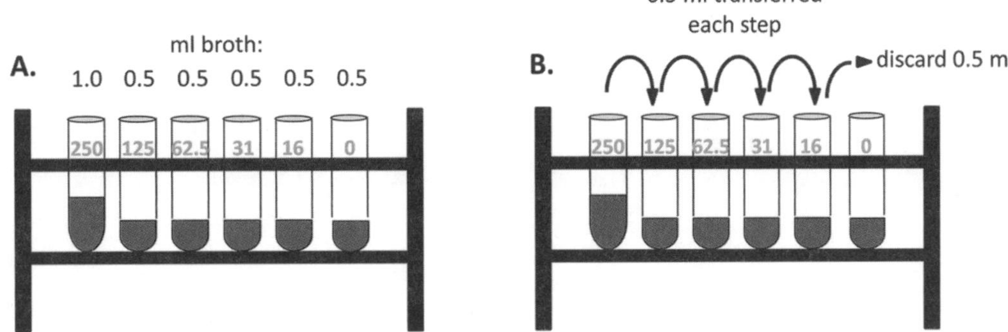

Figure 3-16 Preparing dilutions of antibiotic for the broth tube MIC assay.

PROCEDURE

Setting up a MIC dilution assay

Working in pairs, select one antibiotic for testing.

Step one: Preparing tubes containing medium with different amounts of antibiotic so that the final concentrations will be 16 to 250 µg per ml in 2-fold steps.

1. Set up a series of six capped, sterile culture tubes and label them 250, 125, 62.5, 31, 16, and 0.
2. Add 1 ml broth to tube "250" and 0.5 ml to the other tubes (Fig. 3-16A).
3. Add antibiotic from the stock solution to tube "250" so that the concentration is 500 µg (0.5 mg) per ml. The concentration of antibiotic in each stock solution is 25 mg per ml. The amount to be added is calculated by:

$$25\ (\text{mg/ml}) \times V\ (\text{ml}) = 0.5\ (\text{mg/ml}) \times 1\ (\text{ml}) = 0.02\ \text{ml} = 20\ \mu\text{l}$$

   This is just another example of using the equation $C_1 \times V_1 = C_2 \times V_2$ (Challenge Two).
4. Gently shake tube "250" or vortex so that the concentration of antibiotic is uniform, then transfer 0.5 ml to tube "125." Shake and transfer 0.5 ml to tube "62.5" and continue until 0.5 ml from tube "32" is added to tube "16." Remove 0.5 ml from tube "16" and *discard* (Fig. 3-16B). When you are done, you will have six tubes containing 0.5 ml broth, five tubes with *twice* the final concentration of antibiotic, and one tube with no antibiotic.

Step two: Diluting cells to a concentration of 1×10^5 cells/ml in each tube.

Each tube must contain not only the correct final concentration of antibiotic but also 10^5 viable cells. This can be achieved by adding 0.5 ml broth containing 10^5 cells to each tube. For this addition, you need cells diluted in broth to a final concentration of 2×10^5 viable cells/ml.

Your instructor will provide the class with exponential-phase cells in broth and will tell you the OD_{600}.

Last week you calculated the relationship between OD_{600} and viable cell counts for exponential-phase cells. Using this relationship and serial dilutions, you can prepare a broth culture with the correct concentration of cells.

Procedure continues on next page

Setting up a MIC dilution assay (*continued*)

EXAMPLE

1. Suppose 1 $OD_{600} = 1.4 \times 10^9$ viable cells/ml and you are given a culture with $OD_{600} = 0.265$ (Fig. 3-17). In that case, the concentration of viable cells (C_1) is:

$$\frac{OD_1}{OD_2} = \frac{C_1}{C_2}$$

$$\frac{0.265}{1} = \frac{C_1}{(1.4 \times 10^9)}$$

$$C_1 = 0.265 \times 1.4 \times 10^9 = 0.37 \times 10^9 \text{ or } 3.7 \times 10^8 \text{ cells/ml}$$

2. The concentration of cells is too large to simply withdraw a volume from the culture and add to broth for a concentration of 2×10^5 cells/ml, so the culture must be serially diluted to a more manageable concentration. Two serial dilutions will lower the concentration 100-fold, to 3.7×10^6 cells/ml. You can then use this tube to make the final desired concentration.
3. You will add 0.5 ml of a dilution containing 2×10^5 cells/ml to each of six tubes, so you need 3 ml. You decide to make up 5 ml of cells at this concentration.

   You know the initial concentration of cells (the concentration in the serial dilution tube), the desired final volume, and the desired final concentration. Since

$$\mathbf{C_1 \times V_1 = C_2 \times V_2}:$$

$$3.7 \times 10^6 \text{ (cells/ml)} \times V \text{ (ml)} = 2 \times 10^5 \text{ (cells/ml)} \times 5 \text{ (ml)}$$

$$V = 0.34 \text{ ml}$$

   This result tells us that 0.34 ml of the serially diluted culture contains the number of cells needed for 5 ml of cells with a concentration of 2×10^5 cells/ml.

4. Adding 0.34 ml of the serially diluted culture to 4.66 ml broth will result in 5 ml diluted cells at a concentration of 2×10^5 cells per ml.

Step three: Completing the setup for the MIC assay.

Add 0.5 ml of the cells prepared in step two to each of the dilutions prepared in step one. The addition dilutes both the cells and the antibiotic to the correct final concentration (Fig. 3-18). Incubate in a shaking water bath overnight at 30°C.

Figure 3-17 Diluting cells to the correct concentration for the MIC assay.

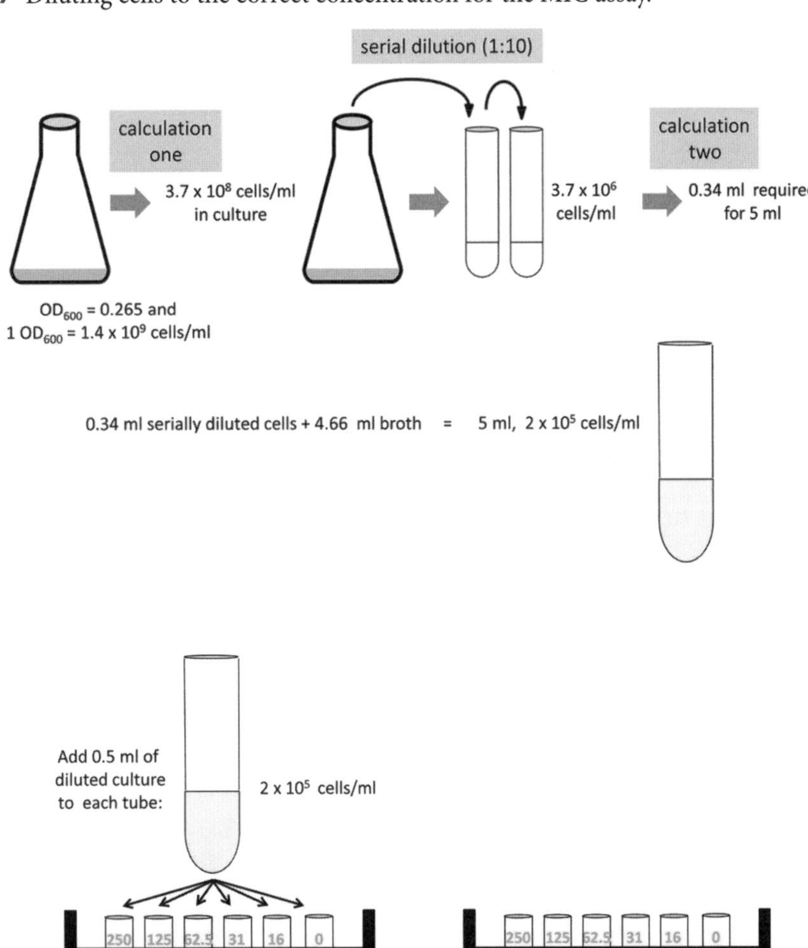

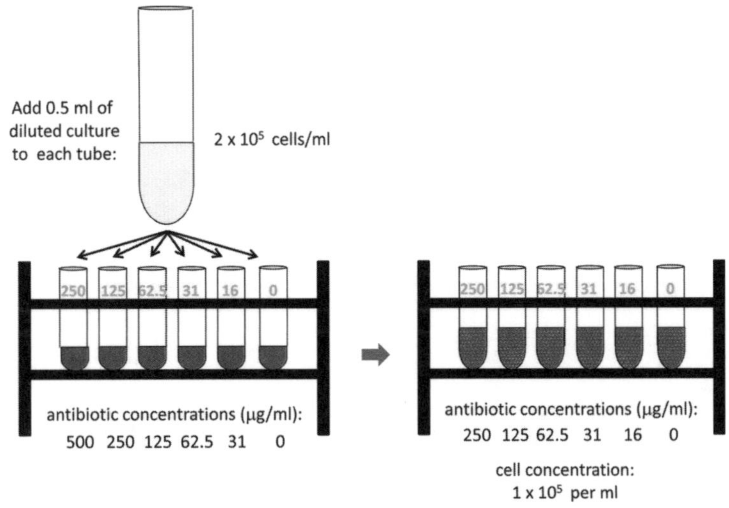

Figure 3-18 Setting up the MIC assay: final step before incubation.

QUESTIONS

1. Transcription, translation, and cell wall synthesis are common targets of antibiotics. Why do you think actively growing cells are more susceptible to antibiotics than quiescent (inactive) cells? (B3)
2. In the example for preparing an inoculum for the MIC assay, what volume of cells from the *undiluted* culture would need to be added to 5 ml to obtain inoculum with the desired concentration? (B3)
3. For the MIC plate assay shown in Fig. 3-14, rank the antibiotics according to their activity against the test strain. (B4)

Solving Challenge Three

1. *Construct a graph of bacterial growth by measuring optical density.*
2. *Determine the concentration of viable cells by serial dilution and plating.*
3. *Calculate the relationship between optical density and viable counts for exponential-phase cells.*
4. *Determine the minimal inhibitory concentrations of different antibiotics by 2-fold dilutions in broth medium.*
5. ***Interpret the results to identify the antibiotics that might be useful for treatment.***

Add your results to the appropriate row in Table 3-4 and determine the MIC for each antibiotic.

+ = turbidity similar to control (no antibiotic)

+/– = reduced turbidity compared to control

– = no turbidity (broth is clear)

Using the data in Table 3-4, determine which antibiotics would be potentially useful in treating patients infected with the *Pseudomonas* strain. Would it be worthwhile to test any of these antibiotics at a concentration lower than 16 μg/ml?

Table 3-4 Results of MIC assay

Antibiotic	Initials	Antibiotic concentration (μg/ml)						MIC
		250	125	62.5	31	16	0	
ampicillin								
chloramphenicol								
kanamycin								
nalidixic acid								

BIBLIOGRAPHY

Arnott ID, Landers CJ, Nimmo EJ, Drummond HE, Smith BK, Targan SR, Satsangi J. 2004. Sero-reactivity to microbial components in Crohn's disease is associated with disease severity and progression, but not NOD2//CARD15 genotype. *Am J Gastroenterol* **99:**2376–2384. http://dx.doi.org/.

Cano RJ, Borucki MK. 1995. Revival and identification of bacterial spores in 25- to 40-million-year-old Dominican amber. *Science* **268:**1060–1064. http://science.sciencemag.org/content/268/5213/1060.abstract.

Dubos RJ, Middlebrook G. 1947. Media for tubercle bacilli. *Am Rev Tuberc* **56:**334–345.

Livermore DM. 2002. Multiple mechanisms of antimicrobial resistance in *Pseudomonas aeruginosa*: our worst nightmare? *Clin Infect Dis* **34:**634–640. https://academic.oup.com/cid/article-lookup/doi/10.1086/338782.

Tille PM. 2014. *Bailey & Scott's Diagnostic Microbiology*, 13th ed. Elsevier Mosby, St. Louis, MO.

Whitman WB, Coleman DC, Wiebe WJ. 1998. Prokaryotes: the unseen majority. *Proc Natl Acad Sci U S A* **95:**6578–6583. http://www.ncbi.nlm.nih.gov/pubmed/9618454.

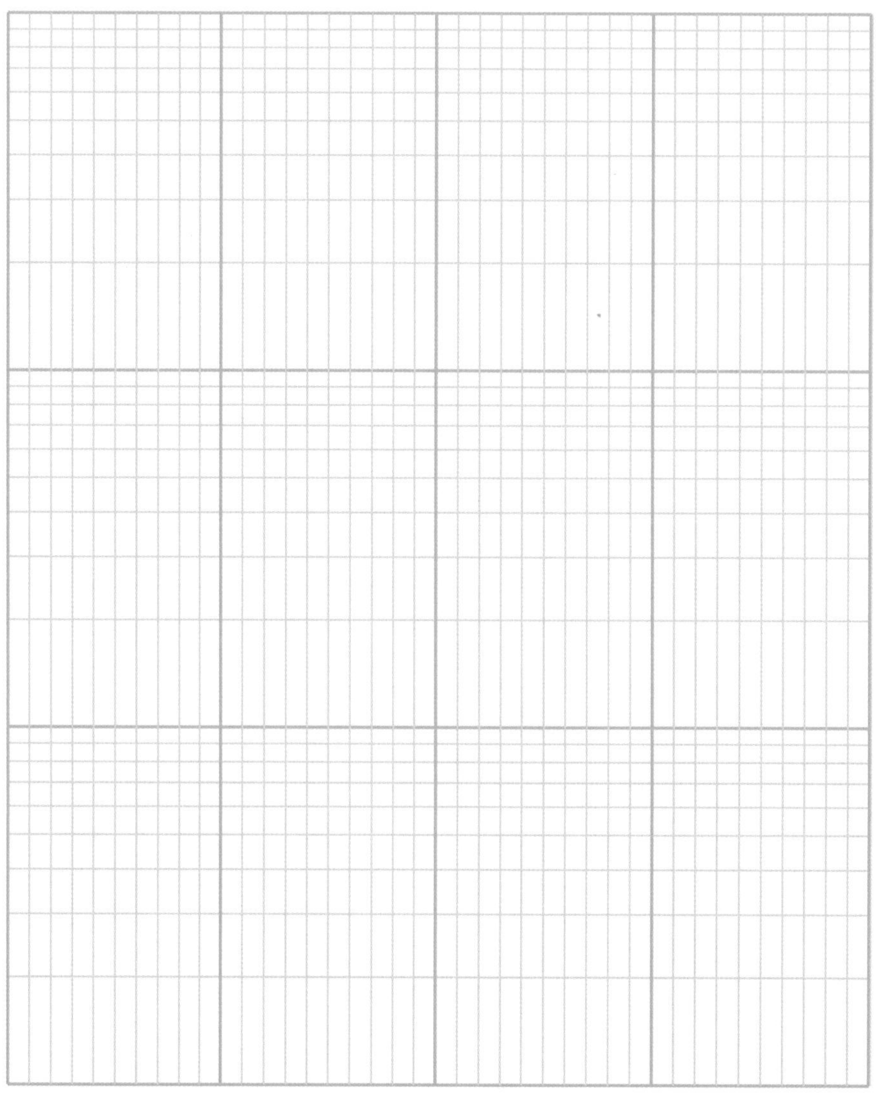

challenge Four

Tracking down the source of an *E. coli* strain causing a local outbreak of disease

Fifteen cases of intestinal illness due to *Escherichia coli* have been reported to the public health department. The pathogenic strain has been isolated from victims and was found to be resistant to chloramphenicol. Health inspectors determined that all the victims had attended a farmer's market the previous weekend and that the only item purchased by everyone was lettuce. There were eight different vendors of lettuce at the market. Lettuce was obtained from each, and *E. coli* resistant to chloramphenicol was isolated from three of the samples. You are asked to determine if one or more of these isolates are the probable cause of the illness and if the resistance to chloramphenicol can be acquired by horizontal gene transfer (HGT).

QUESTIONS BEFORE YOU BEGIN THE CHALLENGE

1. Restate the challenge in the form of two separate hypotheses.
2. Why do you think that the health department would like to know if the chloramphenicol resistance of the pathogenic strain can be acquired by horizontal gene transfer?
3. Why has the health department decided to isolate bacteria from samples of lettuce sold at the market?
4. Suppose it turns out that the bacteria from the lettuce samples are different from the pathogenic strain. Does this rule out the farmer's market as the source of infection? What other scenarios involving produce from the market might be possible?

Strategy for Challenge Four

1. *Isolate plasmid DNA from the pathogenic strain and each of the lettuce isolates.*
2. *Determine for each strain whether chloramphenicol resistance can be transferred by conjugation.*
3. *If the pathogenic strain contains a plasmid, determine if it is related to any of the plasmids from the lettuce isolates.*
4. *Decide if any of the lettuce isolates are the likely source of the pathogenic strain and if the resistance to chloramphenicol can be spread by horizontal gene transfer.*

Lab One

BACKGROUND

Genomic diversity and HGT

KEY POINTS

- Humans and most other animals and plants are genetically diverse because they have different combinations of gene alleles acquired during reproduction.
- In bacteria, genetic diversity is independent of reproduction and in many cases largely due to HGT, the movement of DNA from one cell to another.
- Bacteria can acquire by HGT genes that are completely new to the species. These genes can confer important changes in the phenotype of the cell.

No two individuals are exactly alike. Each of us has a unique set of genetically determined characteristics. Together, these characteristics make up the **phenotype** of an individual. Where does this variability come from? Partly it is derived from the same gene having different mutations. Each gene variant due to mutation is called an **allele**, and different alleles can determine everything from eye color to cholesterol levels to fatal disease. Identical twins have the same alleles for nearly every gene, but even here different mutations occurring in each embryo during development can result in small but distinct differences between the two individuals (Li et al., 2014).

Alleles are essential for phenotypic diversity, but it is the new combinations of these alleles that are largely responsible for the variation between individuals. In humans and most other eukaryotes, the creation of diversity is closely linked to the mechanism of reproduction. Offspring inherit genes from both parents, resulting in an overall genetic makeup that consists of a new set of alleles. In addition, chromosome fragments can be exchanged at meiosis, during the formation of sperm and egg cells. This exchange can result in offspring inheriting a chromosome that itself contains a new combination of alleles. Notice that to a very large extent every person has the same *set* of genes. It is the novel combination of alleles that accounts for most of the differences between individuals.

In bacteria, genetic diversity is not generated during reproduction. The number of individual cells usually increases by simple cell division and results in a clone of cells that are genetically identical and therefore phenotypically the same. For this reason, it used to be thought that rarely-occurring mutations solely accounted for variability. Probably this is true for some bacteria, particularly those highly adapted to a stable and isolated environmental niche. We have come to realize, though, that many bacteria, including a host of important pathogens, have mechanisms that can generate enormous phenotypic variability within a species. It is just that these mechanisms take place independently of reproduction.

In the 1940s, Joshua Lederberg noticed that when certain strains of *E. coli* were mixed together, a strain with a new phenotype appeared. This strain had alleles from *both* of the original strains, indicating that there had been allelic reassortment. It turned out that this reassortment was due to the transfer of DNA from one strain to the other. This phenomenon was a laboratory curiosity (and an important tool for microbial geneticists) but seemed to have little practical relevance otherwise. This was an unfortunate mistake, with consequences that continue to affect us today.

One of the first indications that bacterial gene transfer was important, and indeed had serious implications for human health, came after World War II. At the time, antibiotics were being used to treat bacterial dysentery in Japan. The disease is caused by *Shigella dysenteriae*, a very close relative of *E. coli*. Over a number of years strains arose that were resistant to the antibiotics used in treatment (Watanabe, 1963). This might have been explained by a mutation in a gene, but some strains became resistant to multiple antibiotics simultaneously (Fig. 4-1). Since the antibiotics were unrelated and worked by different mechanisms, attributing these resistances to a single mutation seemed unlikely. It turned out that genes for multiple drug resistances were being transferred from cell to cell not only within a species but between species. These genes were on an extrachromosomal element called a **plasmid**, which can be replicated and maintained in the cell.

The transfer of genes from cell to cell, independent of reproduction, is called **horizontal gene transfer (HGT)**. There are two basic outcomes for HGT. The first is the one observed by Lederberg: new alleles of genes already present in the chromosome are introduced. The second outcome of HGT is that cells can acquire completely *new* genes. This is what happened during the spread of multiple drug resistances in *Shigella*, and has happened many times since, leading to a dramatic loss in the clinical effectiveness of antibiotics over time. In the case of antibiotic resistances, the new genes resulted in a change in phenotype that was relatively simple and easy to understand. However, whole blocks of genes, encoding different and complex properties, can also be transferred by HGT. Genes essential for pathogenesis, which requires a host of properties and interacting networks of genes, can be acquired this way in *E. coli*.

The shifting genome of many bacteria

KEY POINTS

- Many bacteria, including *E. coli*, have a core set of genes that is present in all strains and defines the characteristics of the species.
- In addition to the core genes, bacteria contain genes from a much larger number of available genes. The set of all the genes available to a species is called the pan-genome.
- The large number of genes available to *E. coli* is the basis for the very different phenotypes of different strains within this species.

TABLE 1. ***Statistics of antibiotic-resistant shigellae in Japan****

Year	No. of strains tested	No. of strains resistant to						
		Sm	Tc	Cm	Sm and Cm	Sm and Tc	Cm and Tc	Sm, Cm, and Tc
1953	4,900	5	2	0	0	0	0	0
1954	4,876	11	0	0	0	0	0	0
1955	5,327	4	0	0	0	0	0	1
1956	4,399	8	4	0	0	0	1	0
1957	4,873	13	45	0	2	2	0	37
1958	6,563	18	20	0	7	2	0	193
1959	4,071	16	32	0	71	0	0	74
1960	3,396	29	36	0	61	9	7	308
Total	38,405	104	139	0	141	13	8	613

Figure 4-1 Table showing the increase in frequency of antibiotic-resistant strains of *Shigellae dysenteriae* in the 1950's. From Watanabe (1963). Despite some fluctuation from year to year, the frequency of strains resistant to one or more antibiotics is increasing overall during this time period. Moreover, strains resistant to all three antibiotics show the greatest rate of increase.

When the sequencing of the human genome was completed, the eukaryotic genes identified were assumed to be present in all members of our species, with only rare exceptions. That is, the sequence gave us the set of genes that define us as human, regardless of our origin or genetic characteristics. However, when the first *E. coli* genome was sequenced (Blattner et al., 1997), no one thought that it could be used to represent the genome of all strains of *E. coli*. Indeed, since the first sequence was published, many different strains have been sequenced, and in every case the genome is different. The explanation is that each strain has a different complement of genes acquired by HGT, both from other strains of *E. coli* and from other species as well. It thus makes little sense to talk about a single *E. coli* genome. Instead, microbiologists use the term "**pan-genome**" to describe the set of all the genes available to a species. The "**core genome**" is the set of genes common to all strains within a species and encoding the properties that define the species. For *E. coli*, a recent comparison of many different strains indicated that the core genome consists of about 2,000 genes but the pan-genome is almost 10-fold larger, about 18,000 genes (Touchon et al., 2009). A typical *E. coli* genome is made up of approximately 4,300 to 5,500 genes. Thus, more than half of the genes are selected from the pan-genome and are in addition to the core genes (Fig. 4-2). This remarkable diversity explains why different strains of *E. coli*, both pathogenic and nonpathogenic, can have such a variety of different characteristics.

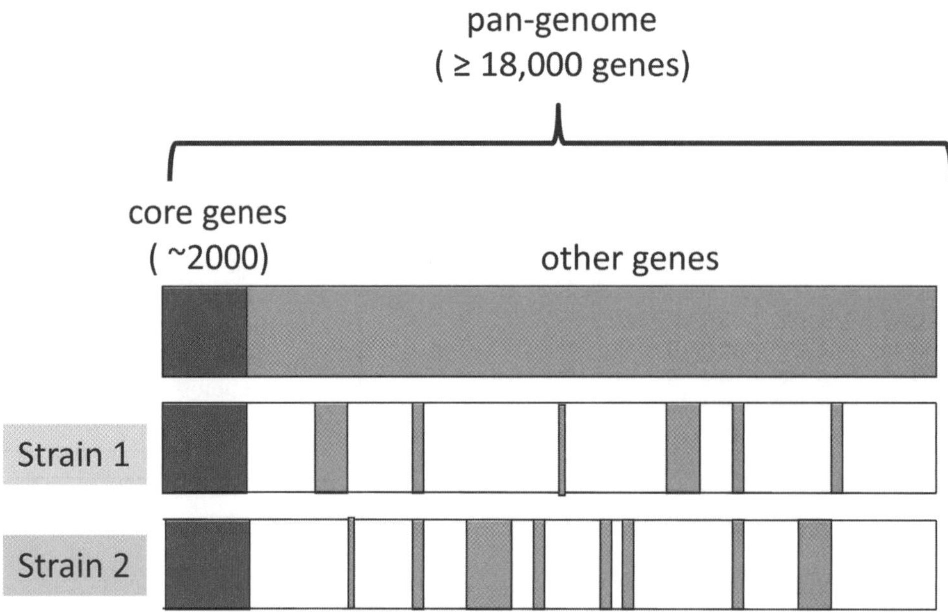

Figure 4-2 Representation of the size of the core- and pan-genomes of *E.coli*. Different strains contain approximately 2,000 core genes, as well as a 2300—3500 genes selected from those in the pan-genome.

In summary, while both eukaryotes and many bacteria have mechanisms for generating diverse phenotypes, there are two fundamental differences.

1. In eukaryotes, the creation of diversity is closely linked to reproduction. In bacteria, reproduction results in cells that are phenotypically and genetically unchanged.
2. Phenotypic differences in eukaryotes mostly reflect the presence of different alleles, with each species having the same set of genes. Phenotypic change in bacteria can involve the organism acquiring not just different alleles but whole new sets of genes.

Conjugation and other mechanisms of HGT

KEY POINTS

- The three principal mechanisms of HGT are transformation, transduction, and conjugation.
- Transformation is limited by the inability of many bacteria to take up DNA and incorporate it into the chromosome.
- Transduction is limited by the narrow range of hosts infected by most bacteriophages.
- Conjugation is a powerful mechanism for HGT because of its broad host range and potential to transfer a large fragment of DNA.
- Plasmids are frequently transferred by conjugation.
- Incoming DNA can be stably inherited if it is on a plasmid able to replicate in the new host or if it is part of a mobile element that can insert into the chromosome or a resident plasmid.

HGT in bacteria can occur by several different mechanisms. The cells of some species can take up DNA in the environment and incorporate it successfully into the genome, a process known as **transformation**. Natural transformability, or **competence**, is widely distributed in the bacterial world, but its occurrence is sporadic and unpredictable (Mell and Redfield, 2014). *Haemophilus influenzae*, a Gram-negative organism, and *Bacillus subtilis*, which is Gram positive, are both naturally competent, although transformation occurs by different mechanisms. *E. coli* is not naturally competent, because it lacks efficient mechanisms for transporting DNA into the cell and incorporating it into the chromosome. However, laboratory techniques that induce a transient ability to take up plasmid DNA or DNA fragments, as well as engineered strains that can integrate this DNA into the chromosome, have been developed for *E. coli* and other species. In another class, you might have had the chance to introduce plasmid DNA into *E. coli* by transformation. If so, you were using cells that had been treated to make them permeable to DNA.

Transduction is another method of HGT. A bacteriophage (or more briefly, phage) is a virus that can infect bacteria (Challenge Five). The phage DNA directs the host machinery to make more phage DNA and the proteins required for new phage particles. The new phages are then assembled from these products and released from the cell. In its simplest form, transduction occurs when a phage particle transports bacterial DNA instead of its own DNA from an infected cell to a new cell. This is due to a mistake in the development of the bacteriophage in the infected cell. At some point during phage assembly phage DNA is packaged into the particle. The developing phage normally distinguishes between its own DNA and the chromosomal DNA of the host, but occasionally a mistake is made and chromosomal DNA is packaged instead. The resulting particle can still infect another cell, but because it contains only bacterial DNA, there is no development of new phages. Instead, the DNA is incorporated into the chromosome of the cell.

HGT by transformation or transduction generally occurs within a narrow range of potential recipients. Many competent strains identify their own DNA and exclude or degrade the DNA from a different species or strain. To initiate infection, transducing phages typically require a specific receptor protein on the surface of the cell. This requirement limits the number of different bacteria that are potential targets for transduction. Finally, incoming DNA, whether introduced by transformation or transduction, must be stabilized in the host, usually by homologous recombination with the chromosome. This requires DNAs with similar base sequences. If the bacteria are not closely related, then there might not be sufficient similarity for homologous recombination.

Probably the most powerful mechanism of HGT is **conjugation**. There are two principal reasons for this. First, a large amount of intact DNA can be transferred at one time, so that whole groups of genes, together encoding complex traits, can be spread from cell to cell. Second, conjugation has a **broad host range**. DNA can be transferred not only within a narrowly restricted group of related bacteria but much more broadly, including between unrelated bacterial species and even into plant and animal cells.

DNA transfer by conjugation requires cell-to-cell contact, and usually the transferred DNA is a plasmid. The DNA is replicated during transfer, with one plasmid copy remaining in the donor cell and the other appearing in the recipient cell. As a result, the proportion of bacterial cells containing the plasmid increases as it spreads throughout a population (Fig. 4-3). The mechanism of plasmid DNA transfer by conjugation in *E. coli* and many other bacteria is outlined in Fig. 4-4A to D. A plasmid-encoded protein, the relaxase, binds to a specific site on the plasmid DNA (A). The relaxase then cleaves one of the DNA strands and in the process becomes covalently linked to the 5' end (B). A multiprotein complex is responsible for transferring the DNA. This complex recognizes the relaxase and secretes it into another cell (C). During secretion, the attached DNA strand is unwound from its complement and, still attached to the relaxase, transferred as well. Transfer probably occurs through a hollow appendage, the pilus. After entry into the recipient cell, the two ends of the DNA strand are rejoined by the relaxase (D). During or just after transfer, the transferred strand is replaced in the donor cell by replication and the complement to the transferred strand is synthesized in the recipient cell.

If new genes introduced by HGT are going to permanently change the phenotype of a cell, the DNA must be faithfully inherited from generation to generation. As indicated above, stabilizing this DNA by homologous recombination with the chromosome requires regions of shared DNA sequence. This is unlikely if the incoming

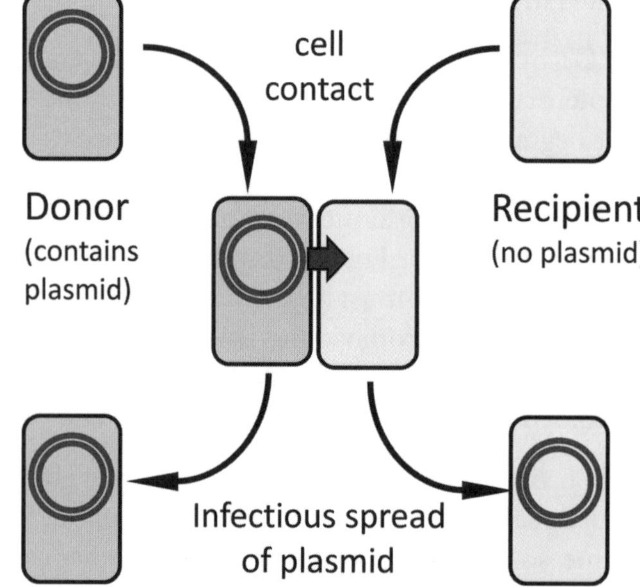

Figure 4-3 Conjugation involves DNA replication and the consequent spread of plasmid DNA.

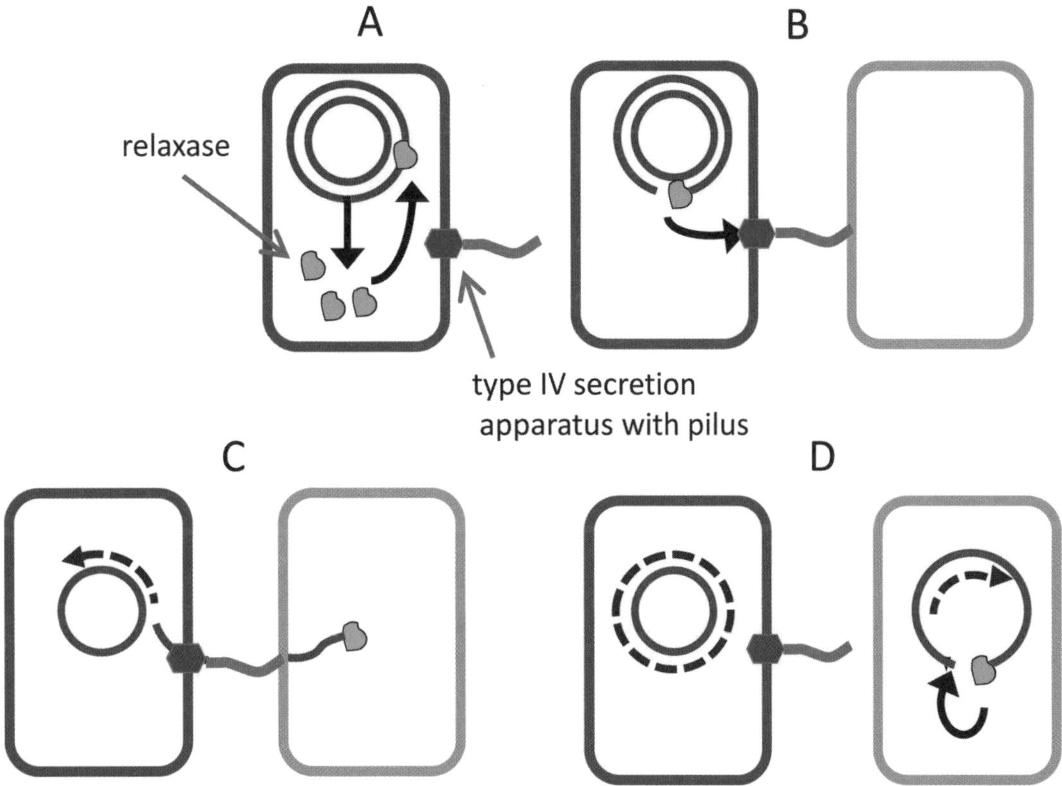

Figure 4-4 Steps in the transfer of plasmid DNA by conjugation. Dashed lines are the newly-synthesized DNA strands.

DNA is from a distantly related species or if entirely new genes are being introduced. Other mechanisms for the stable inheritance of incoming DNA are therefore important for successful HGT. One of these is stabilization by replication. In this case, DNA containing the new genes is part of a plasmid that can replicate in different bacteria. Copies of plasmid DNA are then faithfully passed into daughter cells during cell division. Much of the DNA transferred by conjugation is in the form of a plasmid, and some of these plasmids are able to replicate in many different bacteria.

DNA can also be stabilized by mobile DNA elements. These elements, and there are quite a lot of them in bacteria, are defined segments of DNA that can move to a new location by a mechanism other than homologous recombination. In many cases, the mobile segment simply excises from one site and then inserts itself into another site. When DNA containing foreign genes becomes linked to these elements, then this DNA is moved as well. If the DNA inserts into the chromosome or a plasmid in the host, then the foreign genes become stabilized. Single genes, such as those encoding resistance to an antibiotic, are often located within mobile elements. Huge segments of DNA, containing foreign genes encoding complex properties such as pathogenesis, have also been introduced into the chromosome in much the same way.

Lab One

1. ***Isolate plasmid DNA from the pathogenic strain and each of the lettuce isolates.***
2. ***Determine for each strain whether chloramphenicol resistance can be transferred by conjugation.***
3. *If the pathogenic strain contains a plasmid, determine if it is related to any of the plasmids from the lettuce isolates.*
4. *Decide if any of the lettuce isolates are the likely source of the pathogenic strain and if the resistance to chloramphenicol can be spread by horizontal gene transfer.*

Learning outcomes

After this lab, students will be able to:

a. Set up a conjugation experiment.

b. Describe the steps of the plasmid isolation protocol.

I. Determine if chloramphenicol resistance can be transferred by conjugation

The pathogenic strain of *E. coli* and the three strains obtained from lettuce samples are all resistant to chloramphenicol. Are any of these able to transfer by conjugation the gene for chloramphenicol resistance? In other words, can chloramphenicol-sensitive cells be converted to resistant cells by cell-to-cell contact? Simply mixing resistant and sensitive cells and plating on medium containing chloramphenicol will not give us the answer. After colonies of resistant cells appear on the plate, you would not know whether they are all simply colonies of those cells already resistant to the antibiotic. You need a way to identify those cells that have been converted to chloramphenicol resistance. The simplest way of doing this is by having a counterselection to eliminate the cells that were resistant at the start of the experiment but not those arising by the conversion of sensitive cells. For this reason, you will use a chloramphenicol-sensitive strain that is resistant to the antibiotic nalidixic acid. This resistance is due to a mutation in a chromosomal gene and cannot be transferred to the chloramphenicol-resistant strains by conjugation. If chloramphenicol resistance can be spread by conjugation, then it will be acquired by some of the nalidixic acid-resistant cells during the time the two strains are in contact (Fig. 4-5), resulting in cells that are resistant to both antibiotics. These cells can be selected from the mixed population by plating on medium containing chloramphenicol *and* nalidixic acid. The chloramphenicol-resistant cells used in the experiment will not grow, because they are killed by nalidixic acid. The nalidixic acid-resistant cells will only grow if they have acquired resistance to chloramphenicol.

If the chloramphenicol resistance is transferred, the gene is likely to be on a plasmid. If no growth is observed on the plate containing both antibiotics, then the gene might be in the chromosome or on a plasmid that is not transferred by conjugation.

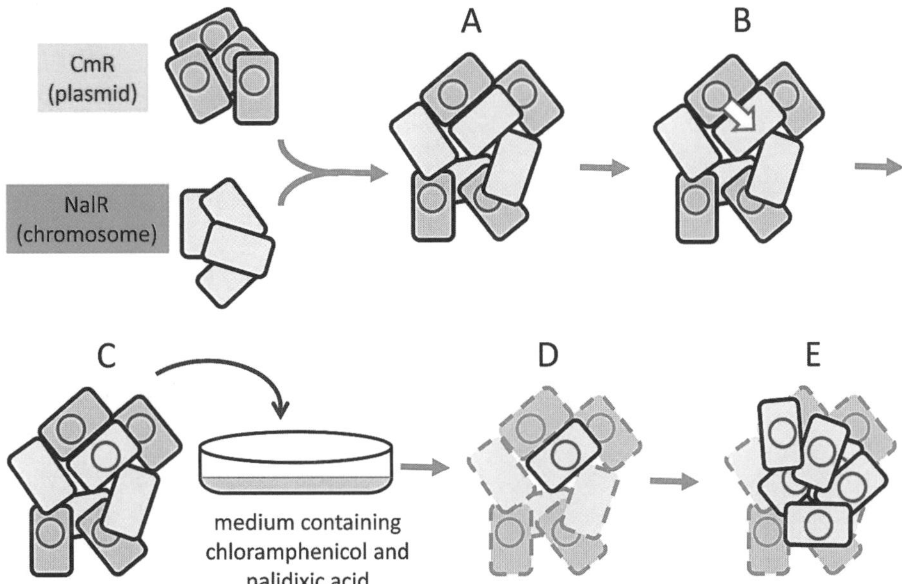

Figure 4-5 Selecting transconjugant cells after conjugation. The donor strain contains a plasmid encoding resistance to chloramphenicol (CmR) and the recipient strain is resistant to nalidixic acid (NalR). (A) Cells of both strains are brought into contact on TSA medium. (B) If the CmR gene is on a plasmid, then the plasmid might be transferred to some of the NalR cells, resulting in cells resistant to both antibiotics. (C) Part of the cell mixture is transferred to TSA medium containing both chloramphenicol and nalidixic acid. (D, E) Cells resistant only to chloramphenicol, and those resistant only to nalidixic acid, are killed. If the NalR cells can acquire CmR by transfer of plasmid DNA, then they will become resistant to both antibiotics and able to grow on this medium.

PROCEDURE

Doing a conjugation experiment on TSA medium

Start the conjugation experiment at the beginning of the lab period so that there will be enough time for the transfer of the DNA.

The following liquid cultures will be available in the lab. Each chloramphenicol-resistant (CmR) culture will have a different code name.

1. The three CmR *E. coli* strains from the lettuce samples.
2. The CmR pathogenic strain isolated from the ill patients.
3. A nalidixic acid-resistant (NalR) laboratory strain of *E. coli*.

(Work singly or in pairs.)

1. At the beginning of the lab meeting, select one of the coded CmR strains for testing. Be sure to write down the name of the strain you will be using. Each of the CmR strains should be tested by approximately the same number of people. You will sample your CmR culture with a loop for the conjugation experiment, and later withdraw 1.4 ml of the culture to isolate plasmid DNA for characterization (below and next session).

Procedure continues on next page

Doing a conjugation experiment on TSA medium (continued)

2. On the back of a plate containing tryptic soy agar (TSA) medium, draw two arrows as shown in Fig. 4-6A. Using a sterile loop, obtain a loopful of cells from the culture of the NalR strain and streak in a line starting from one of the arrows (Fig. 4-6B). Label the streak.

   If it did not seem like there was much liquid on the loop, or if you "skipped" with your loop so that cells were not deposited uniformly along the streak, sterilize the loop and apply an additional loopful of cells.

3. Starting from the second arrow, repeat the previous step for the CmR culture (Fig. 4-6C). It is important to streak with the loop only once and in the direction away from the arrow, not back and forth.

   You can again add additional cells to the streak, but make sure you sterilize the loop first before sampling and again streak only once in the direction of the arrow.

4. Incubate the TSA plate (Fig. 4-6D) at 37°C for about 2.5 hours.

   You will not see a lot of growth during this short time.

5. At the end of the incubation period, circle three areas on the conjugation plate (Fig. 4-6E) and label as shown.

   *Area "CmR" will contain cells of the CmR strain you are testing. These cells will be sensitive to nalidixic acid. Area "NalR" will contain cells resistant only to nalidixic acid. Area "CmR × NalR" will contain both types of cells and, if resistance to chloramphenicol is transferred by conjugation, an additional, third class of cells resistant to both antibiotics. If these **transconjugant** cells are present, they will grow on medium containing both chloramphenicol and nalidixic acid.*

6. Obtain a TSA plate containing nalidixic acid *and* chloramphenicol and divide into three sectors by marking the back as shown in Fig. 4-7 (left). Using a sterile loop, separately sample the cells from the "CmR," "NalR," and "CmR × NalR" areas of the conjugation plate (red circles, Fig. 4-6). Within each area move the loop in a circular motion to gather as many cells as possible, then apply each sample to the appropriate sector on the double antibiotic plate. Start near the middle of the plate and streak back and forth toward the edge of the plate as shown in Fig. 4-7.

7. Incubate the plates approximately 18 to 24 hours at 37°C. After this time, you can return to view the plates or your instructor may store them for you to examine later.

8. Observe the plate for growth. Record your results in Table 4-1, posted online or in the lab.

   The appearance of colonies only in the CmR × NalR sector (Fig. 4-7) means that chloramphenicol resistance was acquired by the NalR strain during incubation on the TSA plate.

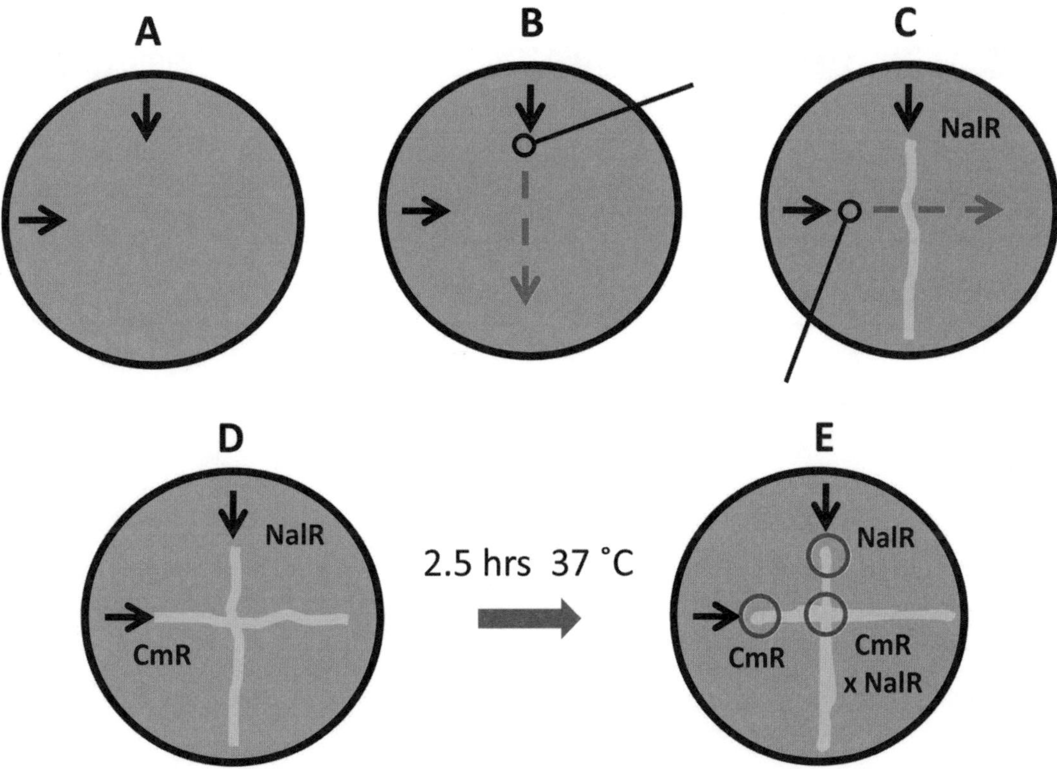

Figure 4-6 Setting up the conjugation experiment on a TSA plate.

Figure 4-7 Testing for the presence of transconjugants on medium containing chloramphenicol and nalidixic acid.

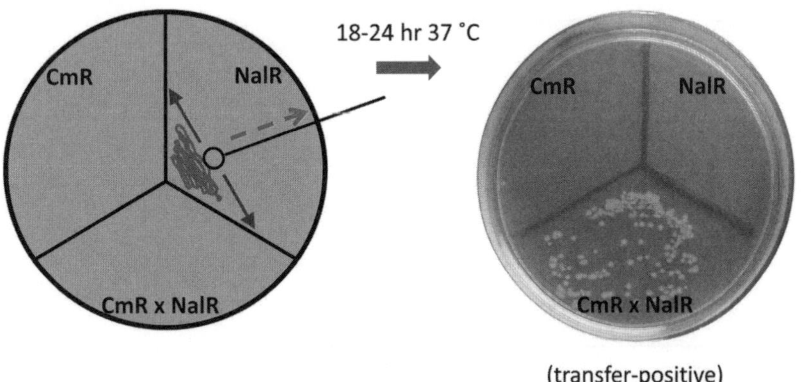

Table 4-1 Results of conjugation experiment

Name(s):	Code name of CmR strain:	Conjugation result (+ or –):

II. Determine if the donor strain for conjugation contains a plasmid

Antibiotic resistance genes are frequently on plasmids that can be transferred by conjugation. This week, you will isolate plasmid DNA by a method that is both rapid and yields plasmid DNA of high quality suitable for digestion with restriction enzymes. In the next lab session, the plasmid DNAs will be characterized from the pattern of fragments after digestion with a restriction enzyme. If the pathogenic strain contains a plasmid, you will then determine if the same plasmid is present in one or more of the lettuce isolates. Many different manufacturers provide kits for the rapid isolation of plasmid DNA, but these are all based on the same underlying principles. The steps below are for ZR Plasmid Miniprep™-Classic. If another kit is used, then follow the similar protocol of the manufacturer.

PROCEDURE

Rapid isolation of plasmid DNA

1. Transfer 1.4 ml of the coded culture you used for conjugation into a 1.5-ml microcentrifuge tube. Centrifuge at 13,000 rpm for 1 minute in a microcentrifuge. The cells will form a pellet adhering to the bottom of the tube.

SAFETY Make sure the rotor is balanced. An unbalanced rotor will shorten the life of the centrifuge and possibly result in cracked tubes, with the contents spilled into the rotor or surrounding chamber. If the rotor has a top, be sure it is placed correctly on the rotor. Tops are important for containing spillage but, if improperly positioned, can fly off the rotor during centrifugation.

2. Pour off the supernatant (medium) into the waste container provided. If necessary, shake the tube several times to remove as much of the medium as possible.
3. Using a micropipette, transfer 200 µl of P1 Buffer (pink) to the pelleted cells. Pipette up and down to resuspend the cells. It is very important that the cells are completely resuspended and that there are no clumps: check carefully.

   P1 Buffer contains Tris [2-amino-2-(hydroxymethyl)-1,3-propanediol] and EDTA (ethylenediaminetetraacetate). The EDTA protects the DNA from nonspecific degradation by nucleases. EDTA chelates (strongly binds to) Mg^{2+}, decreasing its availability as a cofactor for these nucleases.

   Clumped cells are resistant to lysis, and the yield of plasmid DNA will be reduced.
4. Add 200 µl of P2 Buffer (blue) to the cells. Invert gently several times until solution is uniformly purple and viscous. Do not vortex or pipette up and down. Proceed to step 3 *within 2 minutes.*

   P2 Buffer contains a detergent (sodium dodecyl sulfate) and NaOH. This lyses the cells almost immediately. The solution loses turbidity because of the lysis, and the released chromosomal DNA increases its viscosity. In addition, the alkaline pH disrupts base-pairing, thus denaturing the DNA and allowing the complementary strands to separate. For plasmid DNAs, the denaturation is reversible upon neutralization of the solution. However, if the incubation time is more than 2 minutes, the plasmid DNA may become irreversibly denatured and resistant to digestion with restriction endonucleases.

Procedure continues on next page

Rapid isolation of plasmid DNA (continued)

5. Add 400 µl P3 Buffer (yellow). This solution is stored at 4°C (refrigerator). Mix by inverting the tube until the solution is uniformly yellow. Do not mix by pipetting up and down. Let sit at room temperature for about 2 minutes.

   P3 Buffer contains a high concentration of salt and ribonuclease, which digests the RNA. The salt neutralizes the alkaline pH caused by the NaOH, allowing the plasmid DNA to renature. In addition, the salt precipitates the detergent. This precipitate traps the chromosomal DNA. Plasmid DNA is much smaller and is not entrained within the precipitate. If the released chromosomal DNA is handled roughly (e.g., by vortexing), then it is sheared into small pieces that, like plasmid DNA, will remain in the supernatant. This DNA then reduces the purity of your plasmid DNA.

6. Centrifuge the tube at 13,000 rpm for 4 minutes. Before starting this step, it is OK to wait until several other groups are ready to add their tubes or for a microcentrifuge to become available.

   Centrifugation brings down the precipitate, which contains the chromosomal DNA, to the bottom of the tube. The plasmid DNA is not precipitated and remains in the supernatant.

7. While the tubes are spinning, obtain one spin column with its collection tube for each DNA isolation. Be sure to label your column. Place the column in the collection tube.

8. At the end of the centrifugation, pour the supernatant into the column. Pour only once: if there is still some clear liquid at the bottom of the centrifuge tube, do not attempt to collect it. Discard the microcentrifuge tube.

   The supernatant contains plasmid DNA, proteins, carbohydrates, and small molecules.

The following steps are illustrated in Fig. 4-8.

9. Centrifuge the spin column in its tube at 13,000 rpm for 30 seconds.

   The DNA will bind to the column matrix (the small white plug at the bottom of the column). Any undigested RNA will also bind to the column. Since there is much more RNA in the cell than DNA, it would bind all the available sites on the column and prevent the DNA from binding. This is why it is important to degrade the RNA with ribonuclease (step 5). Proteins, carbohydrates, and small molecules do not bind to the column.

10. Remove the column and pour off the liquid in the collection tube. Use the discard jar provided for this and subsequent liquid waste. Replace the column and add 200 µl Endo-Wash Buffer to the column and centrifuge at 13,000 rpm for 30 seconds. In this and subsequent steps, make sure that the tip of the column extending into the collection tube does not come into contact with the flowthrough after centrifugation.

    Endo-Wash inactivates and removes endonuclease I, a major deoxyribonuclease produced by E. coli. *This enzyme binds to the DNA and will degrade it later if it is not removed at this step.*

11. Again pour off and discard the flowthrough. Add 400 µl Plasmid Wash Buffer. Centrifuge at 13,000 rpm for 30 seconds. Make sure the wash solution is tightly capped after use.

    Plasmid Wash Buffer removes any impurities remaining on the column.

12. Pour off and discard the flowthrough. Place the column back into collection tube, and then in the centrifuge. Centrifuge at 13,000 rpm for 1 minute.

This step removes residual wash buffer from the column. The buffer contains ethanol, which can prevent the elution of DNA from the column in the next step. In addition, ethanol inhibits many enzymes, including restriction enzymes, that are often used in subsequent steps.

13. Label a new microcentrifuge tube with the code name of the strain, your names, and the date. Place the column in the labeled tube, being careful not to let any of the wash buffer from the previous step splash onto the tip of the column. Add 30 µl DNA Elution Buffer and let sit for about 1 minute. Be sure you are adding the buffer directly onto the top of the column matrix (the white plug). The buffer will be absorbed into the matrix.

The DNA bound to the column will dissolve in the elution buffer, which has a pH = 8.0. Successful elution of the DNA requires a slightly basic pH.

14. Centrifuge at 13,000 rpm for 30 seconds. Save the microcentrifuge tube, which will contain plasmid DNA dissolved in the elution buffer. Store the tube in the freezer in the designated rack.

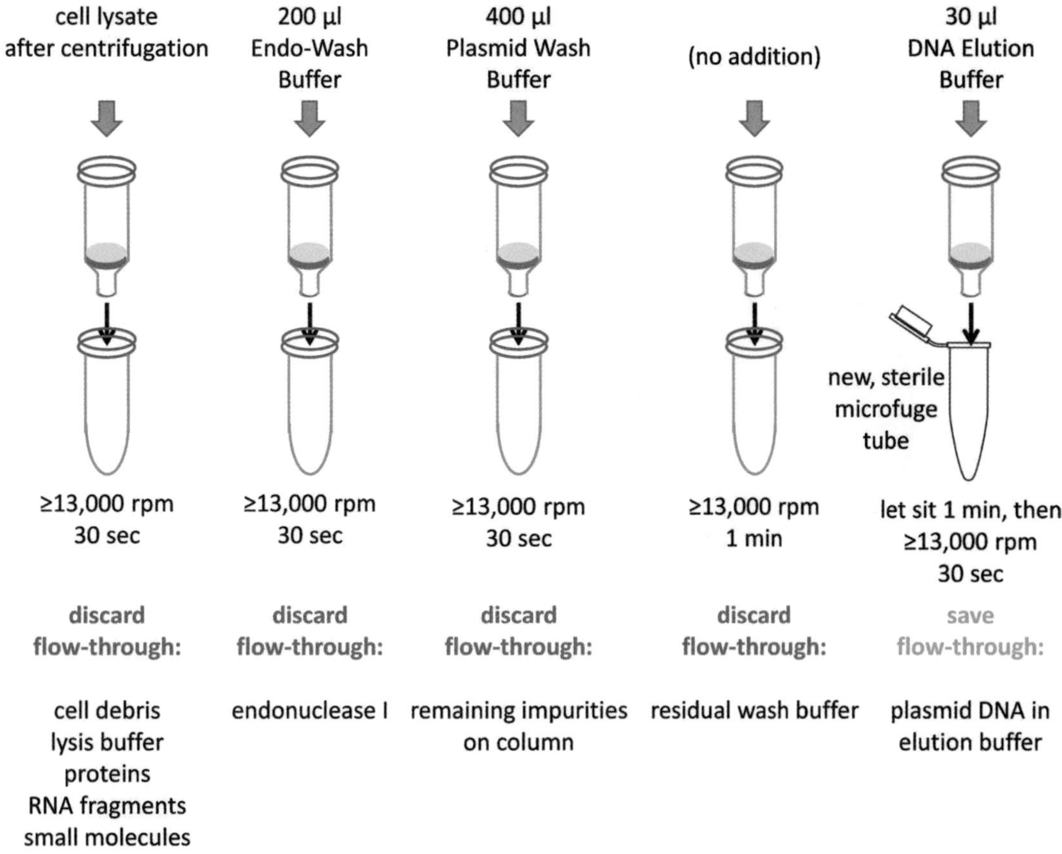

Figure 4-8 Using a spin column to isolate plasmid DNA. Details are for ZR Plasmid Miniprep™ - Classic columns according to the protocol provided by the manufacturer. Steps for the isolation of plasmid DNA with columns from other manufacturers are generally similar. Credit: microfuge image from clker.com.

QUESTIONS

Questions are designated B1 to B6 according to the six levels of Bloom's taxonomy.

1. In the conjugation experiment, why is it important that the gene for NalR is not transferable by conjugation? Think about the results you would get if (a) both resistances are transferable and (b) NalR is transferable but CmR is not. (B2)
2. When you applied the second streak to the conjugation plate, it was important to streak each loopful of culture only once and in the same direction. What would be the problem if you streaked back and forth? (B2)
3. If there are colonies in the CmR × NalR sector, some might argue that this is due to mutation (for example, a CmR cell having a mutation to nalidixic acid resistance and therefore able to form a colony on medium with both antibiotics). What is the argument against this? (B3)
4. Many of the workhorse laboratory strains of *E. coli* contain a mutation completely inactivating endonuclease I. If you were using one of these strains, how might you modify the Miniprep protocol for plasmid DNA? (B3)
5. Strain "AC" of *E. coli* is resistant to ampicillin and chloramphenicol. Strain "KN" is resistant to kanamycin and nalidixic acid. You decide to test whether any of the drug resistances are transferable by conjugation. These strains were streaked onto a TSA plate (Fig. Q4-1) and incubated at 37°C. Cells from within the red circle were then streaked onto TSA containing different combinations of antibiotics. From the results (Fig. Q4-1), can you identify which resistances are transferable? (B5)

Figure Q4-1 Results of conjugation experiment with strains "AC" and "KN".

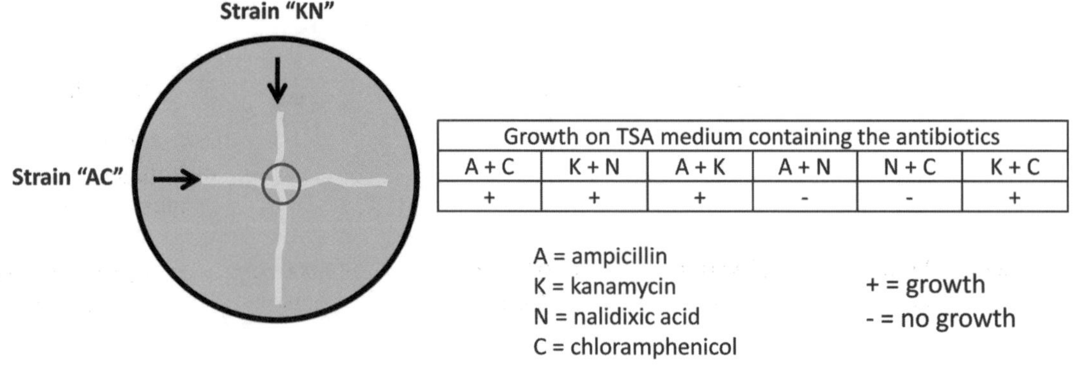

Growth on TSA medium containing the antibiotics					
A + C	K + N	A + K	A + N	N + C	K + C
+	+	+	-	-	+

Lab Two

BACKGROUND

KEY POINTS

- Microbiologists can identify and track different strains of bacteria by the content of their pan-genomic DNA.
- Plasmids are easily separated from chromosomal DNA and for this reason make up a class of pan-genomic DNA that is relatively easy to characterize.
- Digestion with restriction enzymes and separation of the DNA fragments by gel electrophoresis are used to characterize plasmids.
- If plasmids from different strains have a similar pattern of DNA fragments, then these plasmids are likely to be closely related.
- Pulsed-field gel electrophoresis (PFGE) is a technique to resolve very large DNA fragments during electrophoresis. PFGE therefore allows the comparison of total DNA from different strains.
- Current methods of DNA sequencing are becoming practical for the comparison of DNAs from different strains at the level of the nucleotide.

Strain typing

Although it is the core genome that defines a species, the additional genes selected from the pan-genome are often the most interesting, especially to medical microbiologists. For bacteria as diverse as *E. coli* and *Staphylococcus aureus*, these genes are key in determining the degree and mechanism of pathogenicity, the level of resistance to different antibiotics, and the potential for HGT. It is therefore important to identify and characterize these genes. There are many different strains of *E. coli*, each with its own sampling of genes from the pan-genome and with a different potential for causing disease. The pan-genome is not a fixed collection: there are probably genes entering the pool from different species all the time. In addition, new strains appear as the pan-genome is sampled and resampled.

The success of different strains is determined by selection. Thus, in an environment where antibiotics are present, it is the resistant strains that predominate. Similarly, if a particular combination of genes allows a strain to persist longer in a host, it too will have a selective advantage and become more prevalent. If persistence in the host results in disease, then the responsible genes become a matter of medical concern.

Distinguishing between different strains is important to epidemiologists. If there is an outbreak of disease, is it due to a previously unrecognized strain? Does the geographical distribution of the strain allow its source to be identified? Classically, different strains were identified from a small set of variable characteristics, such as the antigenic properties of the outer membrane or the ability to lyse red blood cells. For example, *E. coli* has been characterized by two variable surface structures recognized by our immune system. The first is the O antigen, which is the polysaccharide extending from the outer membrane, and the second is the H antigen, the major component of flagella. Together, the O and H antigens make up the basic serotype of the strain.

Characterization according to serotype or other variable properties continues to be useful, but the method is low-resolution, and as a result, different strains can have the same serotype yet be quite different in other ways. It became clear that characterizing a strain according to its complement of genes from the pan-genome would be a good approach. In the past 30 to 40 years, different methods, with increasingly higher resolution, have been used to do just that.

The genes on plasmids are the easiest part of the pan-genome to characterize. Plasmid DNA can be separated from chromosomal DNA during agarose gel electrophoresis and the size of the plasmid estimated by its rate of migration through the gel. If two isolates contain plasmids of the same size, then it becomes more likely that they are related. This approach was used during an outbreak of *Salmonella* poisoning in 1981 (Taylor et al., 1982). Usually, such outbreaks are localized and easily traceable to a contaminated source, such as the potato salad at a company picnic. Here, the outbreak occurred nearly simultaneously in different states and the victims differed in age (1 month to 73 years, although most were 20 to 29 years) and economic status. Overall, they seemed to have nothing in common. After much sleuthing, it turned out that all were exposed to or had used marijuana, and indeed, a sample was found that was heavily contaminated with *Salmonella*. Was the marijuana from the same primary source? Probably so, since the bacteria isolated from victims of the outbreak all had the same-sized plasmid (Fig. 4-9A).

With the advent of restriction enzymes, plasmid typing at higher resolution became possible. A restriction enzyme cleaves DNA at or near its specific recognition site, which is a particular base sequence or set of sequences. How many times the DNA is cut, and the sizes of the resulting fragments, is determined by the distribution of these sites. If two separately isolated plasmids are closely related, then they will have nearly the same sequence, and as a result, the distribution of restriction enzyme sites along the DNA molecule will be nearly the same. Cleavage of each plasmid with a restriction enzyme will then result in a similar set of fragments, which can be resolved by agarose gel electrophoresis. If the plasmids are not closely related, then the distribution of restriction sites and the fragments obtained after digestion will be different.

In 1982, there was an outbreak of hemorrhagic colitis (essentially, bloody diarrhea) in Oregon and Michigan that was traced to the same fast-food hamburger chain (McDonald's). In both cases, the responsible organism, isolated from many patients, was *E. coli* with what was then considered a rare serotype: O157:H7. The plasmid DNA fragments from the infecting organisms in the two states appeared almost the same after digestion with the restriction enzyme HindIII (can you spot the difference?), suggesting they were closely related (Fig. 4-9B) (Wachsmuth, 1985). *E. coli* from a suspect lot of hamburger meat gave the same plasmid profile as the bacteria isolated from the patients in that state. *E. coli* O157:H7 is now recognized as a serious pathogenic strain causing bloody diarrhea and sometimes life-threatening anemia and kidney failure (hemolytic-uremic syndrome, or HUS). In 1993, there was another, more serious outbreak of food poisoning caused by this strain (below). Cattle are the normal carriers but do not show symptoms. Sources of human infection are the result of contaminated carcasses in the slaughterhouse or plants treated with animal fertilizer.

Figure 4-9 (A) Plasmid content of different *Salmonella* isolates during a multi-state outbreak of *Salmonella* poisoning. The isolates from different patients and the marijuana source all contained the same-sized plasmid (pl), which was easily separated from the chromosomal DNA (chr) by electrophoresis (lanes A – G). Plasmids from *Salmonella* unlinked to the outbreak did not have this plasmid (lanes H and I). From (New England Journal of Medicine, 1982). Credit: Taylor et al. (B) OR, MI: Electrophoresis of HindIII-digested plasmid DNA from two *E. coli* O157:H7 strains isolated in Oregon and Michigan. An unrelated plasmid, isolated from a strain with the same serotype but not involved in the outbreak, is in the third lane (other). From (Infection Control, 1985). Credit: Kaye Wachsmuth.

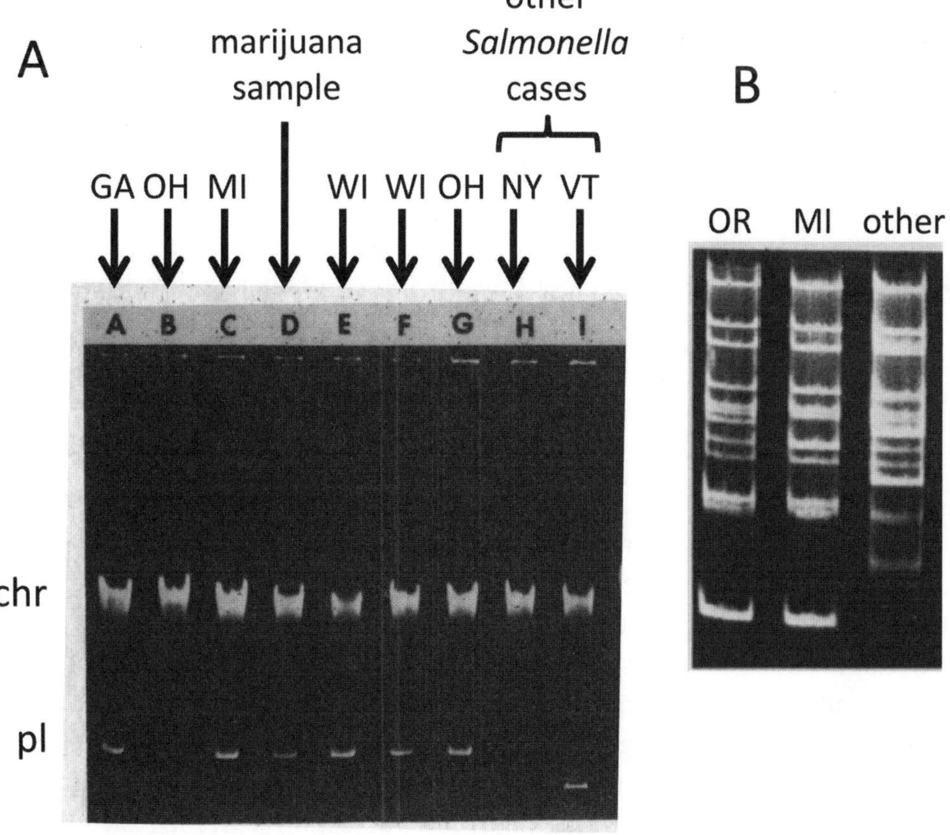

Of course, plasmids account for only a fraction of the genetic differences between strains and in some cases are not present at all. It would clearly be a step forward if the restriction fragments for all the DNA in the cell could be used to determine strain relatedness. The underlying idea is the same: the differences in sequence will be reflected by differences in the pattern of restriction fragments during gel electrophoresis. However, there were two problems with this approach. Most restriction enzymes recognize sites that are four or six base pairs. There are usually many of these sites on the chromosome, so digestion would result in a large collection of fragments that would be impossible to resolve on a gel. Fortunately, a few six-base cutters do give a manageable number of fragments. For example, XbaI cuts the chromosome of MG1655, a standard laboratory strain of *E. coli*, only 35 times. The reason is that the tetranucleotide CTAG is very rare in *E. coli*, and the recognition site for XbaI (TCTAGA) includes this sequence of bases. A more general solution to the problem of too many fragments was the discovery of restriction enzymes having eight-base-pair recognition sites. These sites are far less frequent on the DNA, and therefore digestion with these enzymes leads to a manageable number of fragments. An example is NotI, which has the recognition site GCGGCCGC.

A second problem with analyzing total genomic DNA is that if there is a small number of fragments, many of these will be very large, often several hundred thousand base pairs. Fragments larger than a mere 10,000 to 20,000 base pairs cannot be resolved by ordinary agarose gel electrophoresis. The solution was **pulsed-field gel electrophoresis (PFGE)** (Fig. 4-10). In this technique, fragments are loaded into the well of an agarose gel and a current applied so that the DNA migrates along the length of the gel toward the positive terminal. So far, this is just like ordinary gel electrophoresis, and most of the fragments will migrate at about the same rate because of their large size. In PFGE, additional electric fields, at different angles relative to the usual field, are applied as well, with each field occurring in sequence. Because of these additional fields, molecules are periodically forced to change direction and migrate leftward and rightward relative to their normal migration. The key to PFGE is that smaller molecules are able to reorient to a new field and to begin migrating in the new direction more quickly than large molecules (Fig. 4-10). As a result, they make more progress down the gel (spend more time migrating in the normal direction).

PFGE has been a very useful tool in tracking the origin of strains causing outbreaks of disease. In 1993, there was again a serious outbreak of bloody diarrhea in several western states, with four deaths. The cause was again *E. coli* O157:H7. The victims had all eaten at different outlets of the same fast-food hamburger chain (Jack in the Box), but then so had many other people who did not become sick. Were the cases of illness all due to the same strain, suggesting there was a common source? Was the meat from the hamburger chain the likely source of infection? PFGE provided the answers to these questions. Four isolates from patients living in different states had the same pattern of DNA fragments after digestion with the restriction enzyme XbaI and PFGE (Fig. 4-11, lanes 1-4) (Barrett et al., 1994). In addition, the lot of

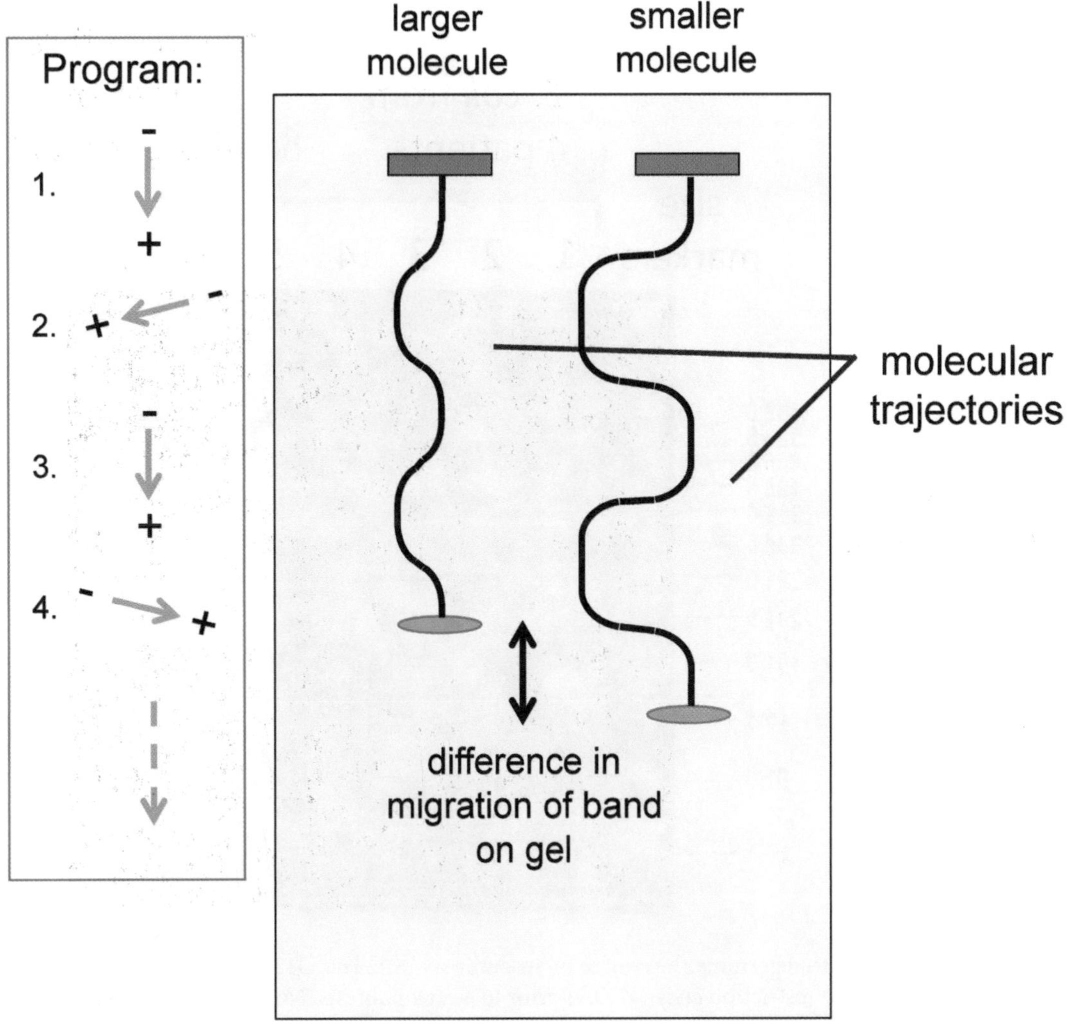

Figure 4-10 Pulsed-field gel electrophoresis (PFGE) with a simple oscillating field.

ground meat sent to each of the restaurants was contaminated with a strain that had DNA with the same pattern (lane 5). *E. coli* from ground meat not implicated in the outbreak (lanes 6 and 7) had different sets of DNA fragments. The people who got sick had probably eaten a hamburger patty that had been undercooked.

In the past several years, whole-genome DNA sequencing has become faster and much less costly, due to the availability of sequencing methods that are "high throughput," meaning that a lot of different DNA fragments can be sequenced at one time. It is now becoming practical to use sequencing as the ultimate tool in tracking strains by their genetic content. In 2011, there was an outbreak of *Salmonella* food poisoning in 44 states, with about 300 people affected. The common food source was traced to a plant that made salami. Isolates were taken from sick individuals, the meat product, several different ingredients used in production, and from a disposal area in the plant itself. When PFGE was used for characterization,

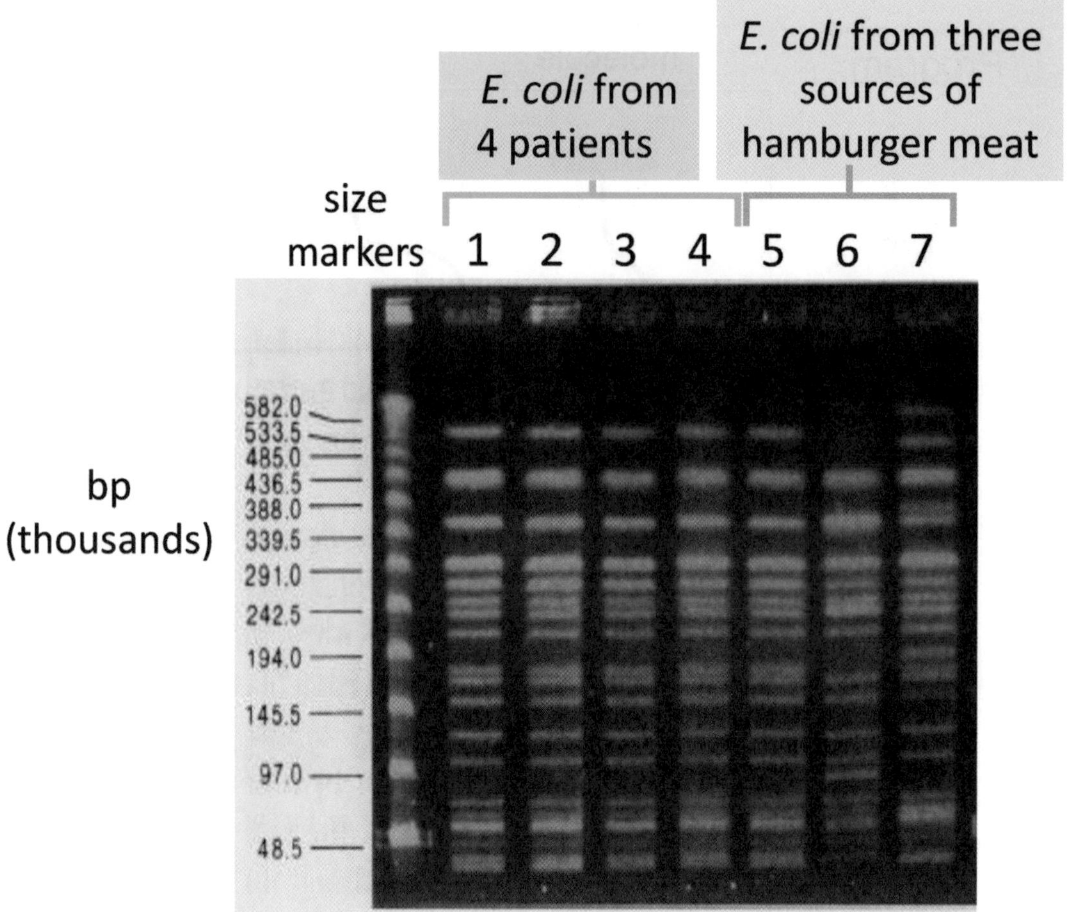

Figure 4-11 PFGE to determine the source of an outbreak of *E. coli* O157:H7. Total DNA in the cell was digested with the restriction enzyme XbaI prior to electrophoresis. Notice the very large size of the DNA fragments. The strain for lane 5 was from meat delivered to the restaurant chain. From *Journal of Clinical Microbiology,* 1994. Credit: (Barrett et al., 1994).

the pattern of DNA bands was unfortunately the same as those from other, unrelated disease-causing isolates. However, when whole-genome sequencing was used, the strain causing the salami outbreak could be easily distinguished from other strains (Lienau et al., 2011). It was found that the red and black pepper used in processing the meat was contaminated, leading to a recall of these ingredients.

Very recently, small, portable DNA sequencers, about the size of a cell phone, started to become available (Fig. 4-12). This will allow sequencing on-site in remote locations where traditional sequencing facilities are unavailable. Although still being developed, the small units have already been used to characterize the Ebola virus from 14 patients in Guinea, West Africa, in a mere 48 hours (Check Hayden, 2015).

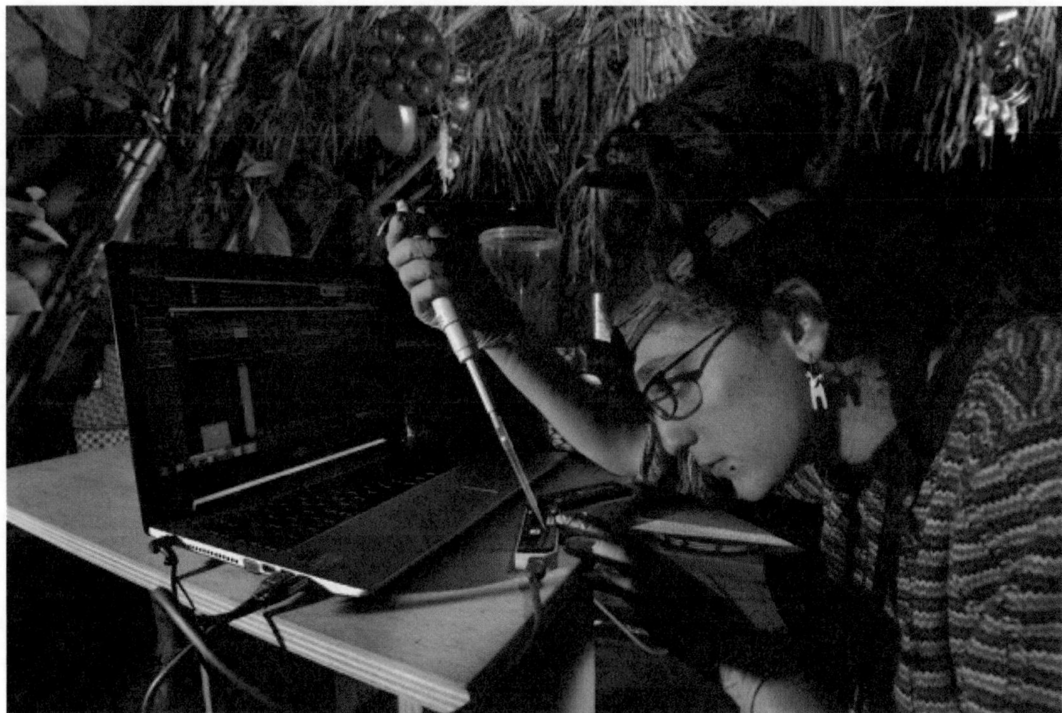

Figure 4-12 A minisequencer in action. Credit: Fabio Pupin/MUSE (Museo delle Scienze di Trento).

Lab Two

1. *Isolate plasmid DNA from the pathogenic strain and each of the lettuce isolates.*
2. *Determine for each strain whether chloramphenicol resistance can be transferred by conjugation.*
3. ***If the pathogenic strain contains a plasmid, determine if it is related to any of the plasmids from the lettuce isolates.***
4. *Decide if any of the lettuce isolates are the likely source of the pathogenic strain and if the resistance to chloramphenicol can be spread by horizontal gene transfer.*

Learning outcomes

After this lab, students will be able to:

a. Set up a restriction enzyme digest.

b. Analyze restriction enzyme fragment patterns to determine plasmid relatedness.

I. Determine if the plasmid DNAs from the lettuce isolate and the pathogenic strain are related

If plasmid DNAs from different bacterial isolates have an identical pattern of bands after digestion with a restriction enzyme and agarose gel electrophoresis, then it is likely they are the same. If some of the plasmid bands are the same, while others are not, then the plasmids are probably related, with each having stretches of DNA with identical or nearly identical base sequences (Fig. 4-9B).

PROCEDURE

Doing a restriction digest

You will digest the plasmid DNA isolated last week with the restriction enzyme SspI. This enzyme was chosen because digestion results in a manageable number of fragments that are easily separated by gel electrophoresis. Each digestion (25 µl) consists of the following components:

____5µl plasmid DNA
______ buffer (provided as a 10× concentrate)
______ deionized H_2O, sterile
____1 µl enzyme SspI
___25 µl total volume

To assemble the reaction mixture:

1. Calculate the amount of buffer and deionized water you will need for the reaction.

   The buffer is 10×, meaning that you use 1 µl for every 10 µl of the final reaction mixture.

   The amount of water is determined by adding up the volumes of all the other components of the reaction, then subtracting this sum from 25 µl.

2. Transfer 5 µl of your DNA to a sterile microcentrifuge tube. Here and below, be careful with your pipetting and make sure you are delivering the correct amount of liquid into the tube.
3. Pipette the calculated amount of sterile H_2O into the tube.
4. Transfer the correct amount of 10× enzyme buffer into the tube.

   Make sure the buffer is completely thawed and mixed before using (you can use a vortexer to mix or invert several times). You can deposit the buffer on the inside of the tube, just above the solution: observing the drop that has been expelled provides assurance that you are adding the buffer.

5. Flick the tube with your finger to mix the contents, and then spin the tube down in a microcentrifuge for a few seconds to bring everything to the bottom.

   This step makes certain that everything is mixed and at the right concentration for the addition of enzyme.

6. Add 1 µl enzyme to the tube. Enzymes are kept in the freezer and lose activity when they become warm. Keep the enzyme on ice during the time you are sampling. Make sure you are pipetting the correct amount and add the enzyme directly into the solution.

   Because the enzyme stocks contain glycerol, if you watch carefully, you can often see the enzyme solution entering the reaction mixture and settling to the bottom of the tube. Do not add too much enzyme: excess enzyme can result in lower specificity and unpredictable results.

7. Mix by flicking and centrifuging as above.

   It is important to mix: since the enzyme solution contains glycerol, it can otherwise remain at the bottom of the tube for long periods.

8. Place the tube in a rack in the 37°C dry incubator. Be sure to label the tube. It is not necessary to use a water bath. Incubate for about 60 minutes.
9. Add 5 µl 6× tracking dye (Challenge Two). Mix by flicking the tube and centrifuge for a few seconds to bring down the drops.

PROCEDURE

Agarose gel electrophoresis of the DNA fragments

SAFETY Ethidium bromide is carcinogenic and mutagenic. Always wear gloves when handling stained gels and solutions containing ethidium bromide. Dispose of solutions containing ethidium bromide properly as described by the instructor. The Tris-borate buffer used in electrophoresis is poisonous when ingested.

Remember:

1. While the enzyme reactions are incubating, prepare agarose gels (Challenge Two) as described by the instructor. If more than one gel is being used, be sure that each gel is numbered.
2. Before loading the gel, make sure that all the CmR isolates will be represented at least once on the gel.

   Size comparisons are much easier and more accurate if all the fragments are on the same gel.
3. Reserve one lane in the gel for 8 µl of the 1-kb ladder (do not load in the exact center of the gel: many imaging systems invert the image, and an off-center marker will allow you to keep track of the location of your samples).
4. Mix your sample thoroughly before applying to the gel. Load your entire sample (30 µl) into a well in one of the gels. Be sure to note which lane contains your sample, the lane that contains marker, and the gel number.
5. After electrophoresis, your instructor will photograph each gel and insert the image into a copy of Table 4-2. Remember, the image might be inverted.

Table 4-2 Agarose gel electrophoresis of DNA fragments after digestion of plasmid DNA with SspI

Gel:________

Name(s):	Strain code:	Lane:	Gel photograph (lane 1 left):
		1	
		2	
		3	
		4	
		5	
		6	
		7	
		8	
		9	
		10	
		11	
		12	
		13	
		14	
		15	

Solving Challenge Four

1. *Isolate plasmid DNA from the pathogenic strain and each of the lettuce isolates.*
2. *Determine for each strain whether chloramphenicol resistance can be transferred by conjugation.*
3. *If the pathogenic strain contains a plasmid, determine if it is related to any of the plasmids from the lettuce isolates.*
4. ***Decide if any of the lettuce isolates are the likely source of the pathogenic strain and if the resistance to chloramphenicol can be spread by horizontal gene transfer.***

When Table 4-2 for your gel is posted, identify your sample and enter your name and the code for your CmR strain in the row for your lane.

Once all the data are posted, the instructor will reveal the identity of the coded strains. From the results, you should be able to answer the following questions:

1. Is the pathogenic strain or any of the CmR isolates from lettuce able to transfer the resistance by conjugation?
2. Do any of these strains contain plasmid DNA?
3. If the pathogenic strain contains a plasmid, is the SspI fragment profile similar or identical to any of the plasmid-containing isolates from the lettuce samples?
4. What is your conclusion about the likely source of the contamination? Remember, it is possible that none of the three isolates from lettuce is the source. It is also possible that more than one of the lettuce samples are contaminated with the pathogenic strain.
5. Estimate the size of any plasmid present: first, estimate the size of each fragment band by comparing to the marker fragments, then add up the sizes.
6. Do you anticipate that the transconjugant strain will contain plasmid DNA? What fragments would be obtained after digesting this DNA with SspI?
7. If more than one plasmid is identified in the different lettuce isolates, do these appear to be related?

BIBLIOGRAPHY

Barrett TJ, Lior H, Green JH, Khakhria R, Wells JG, Bell BP, Greene KD, Lewis J, Griffin PM. 1994. Laboratory investigation of a multistate food-borne outbreak of *Escherichia coli* O157:H7 by using pulsed-field gel electrophoresis and phage typing. *J Clin Microbiol* **32:**3013–3017.

Blattner F, Plunkett G, III, Bloch CA, Perna NT, Burland V, Riley M, Collado-Vides J, Glasner JD, Rode CK, Mayhew GF, Gregor J, Davis NW, Kirkpatrick HA, Goeden MA, Rose DJ, Mau B, Shao Y. 1997. The complete genome sequence of *Escherichia coli* K-12. *Science* **277:**1453–1462.

Check Hayden E. 2015. Pint-sized DNA sequencer impresses first users. *Nature* **521:**15–16. http://www.nature.com/doifinder/10.1038/521015a.

Li R, Montpetit A, Rousseau M, Wu SY, Greenwood CM, Spector TD, Pollak M, Polychronakos C, Richards JB. 2014. Somatic point mutations occurring early in development: a monozygotic twin study. *J Med Genet* **51:**28–34.

Lienau EK, Strain E, Wang C, Zheng J, Ottesen AR, Keys CE, Hammack TS, Musser SM, Brown EW, Allard MW, Cao G, Meng J, Stones R. 2011. Identification of a salmonellosis outbreak by means of molecular sequencing. *N Engl J Med* **364:**981–982.

Mell JC, Redfield RJ. 2014. Natural competence and the evolution of DNA uptake specificity. *J Bacteriol* **196:**1471–1483. http://jb.asm.org/content/196/8/1471.abstract.

Taylor DN, Wachsmuth IK, Shangkuan YH, Schmidt EV, Barrett TJ, Schrader JS, Scherach CS, McGee HB, Feldman RA, Brenner DJ. 1982. Salmonellosis associated with marijuana: a multistate outbreak traced by plasmid fingerprinting. *N Engl J Med* **306:**1249–1253.

Touchon M, Hoede C, Tenaillon O, Barbe V, Baeriswyl S, Bidet P, Bingen E, Bonacorsi S, Bouchier C, Bouvet O, Calteau A, Chiapello H, Clermont O, Cruveiller S, Danchin A, Diard M, Dossat C, Karoui ME, Frapy E, Garry L, Ghigo JM, Gilles AM, Johnson J, Le Bouguénec C, Lescat M, Mangenot S, Martinez-Jéhanne V, Matic I, Nassif X, Oztas S, Petit MA, Pichon C, Rouy Z, Ruf CS, Schneider D, Tourret J, Vacherie B, Vallenet D, Médigue C, Rocha EP, Denamur E. 2009. Organised genome dynamics in the *Escherichia coli* species results in highly diverse adaptive paths. *PLoS Genet* **5:**e1000344. http://dx.doi.org/10.1371/journal.pgen.1000344.

Wachsmuth K. 1985. Genotypic approaches to the diagnosis of bacterial infections: plasmid analyses and gene probes. *Infect Control* **6:**100–109. http://www.jstor.org/stable/30142728.

Watanabe T. 1963. Infective heredity of multiple drug resistance in bacteria. *Bacteriol Rev* **27:**87–115.

challenge Five

Using bacteriophages to identify the farm releasing pathogenic bacteria into a village stream

You are working as a microbiologist in an underdeveloped country. A remote village is experiencing a high frequency of intestinal infection. The village is surrounded by small farms where there are cattle and other animals. You suspect that runoff from the farms is contaminating the stream that provides the village's water supply. It is not clear which farm(s) might be causing the problem, so you collect runoff from the different sites marked on the map of the area (Fig. 5-1). Unfortunately, you cannot culture the microorganisms from the samples because you lack the containment conditions (such as biological safety cabinets and barriers against insects) that are required when working with potential pathogens. Instead, you decide to use the load of *Escherichia coli*-specific bacteriophages in the runoff from each site as a reporter for the level of contamination.

QUESTIONS BEFORE YOU BEGIN THE CHALLENGE

1. What is the hypothesis being tested in this challenge?
2. State the assumptions being made in the approach to solving this challenge.
3. Why does the microbiologist think that animal farms are the most likely source of the contamination?
4. Why are bacteriophages much less hazardous than their potential hosts?

Strategy for Challenge Five

1. *Pass the samples through a filter to purify the phages.*
2. *Titer the flowthrough to determine the concentration of phages in each sample.*
3. *Determine from the titer at each location which farm(s) are a major source of contamination.*

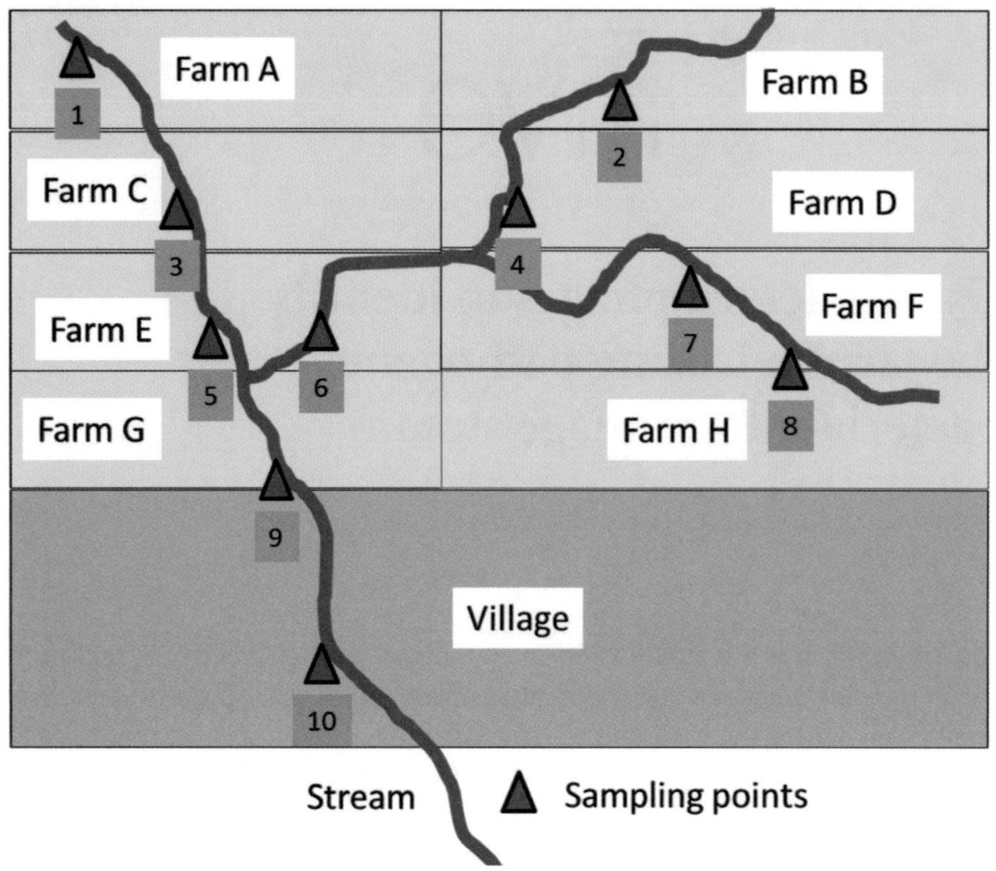

Figure 5-1 Locations where samples were taken from the stream (water is flowing from the farms to the village).

Lab One

BACKGROUND

History and properties of bacteriophages

KEY POINTS

- Bacteriophages (phages) are viruses that infect bacteria. Like all viruses, phages require a host cell for multiplication.
- Bacteriophages are the most common replicating entities on the planet, by far outnumbering bacterial cells and all other viruses.
- The lytic cycle consists of the following sequence: attachment of a phage particle to the cell, entry of phage nucleic acid (DNA or RNA) into the cytoplasm, replication of the phage nucleic acid and synthesis of phage proteins, assembly of new viral particles, and finally their release from the cell.
- The multiplicity of infection (MOI) is the ratio of phage particles to bacterial cells when the culture is first infected.
- If the MOI is low, plaques will form on a lawn of bacteria, with each plaque representing initial infection by a single phage particle.
- When a phage preparation is sufficiently diluted, the number of plaques can be used to calculate the concentration of phage particles.

By 1900 the concept of a virus had become fairly well established, based on landmark experiments with tobacco mosaic disease. This disease results in the unsightly discoloration of tobacco leaves and poor growth of the plant. The blight affects not only tobacco but many other important species as well, such as tomato. The work of Ivanov, Beijerinck, Mayer, and others, reviewed in Lustig and Levine (1992), made clear that the disease was caused by an agent that (i) could pass through a porcelain filter with a pore size so small that bacterial cells were excluded; (ii) was infectious, so that it passed from one plant to another; and (iii) was able to replicate, but only in the presence of host cells. Shortly thereafter, agents with similar properties were found to cause foot-and-mouth disease and yellow fever.

In 1910, the French-Canadian microbiologist Felix d'Herelle was studying a bacterial disease of locusts in Mexico. He observed that when these bacteria were spread out onto agar medium and incubated, there were "clear spots, quite circular, two or three millimetres in diameter, speckling the cultures grown on agar" (cited in Duckworth, 1976). Later, in 1915, F. W. Twort reported an agent that could infect bacteria and was filterable, meaning it could pass through filters that excluded bacterial cells. Two years after that, d'Herelle, now at the Pasteur Institute in Paris, discovered a filterable infectious agent in the feces of patients suffering from bacterial dysentery.

The agent could reproduce, but only in the presence of the bacteria that caused the disease. Moreover, it formed spots on bacterial lawns similar to those he had seen in Mexico. He called this agent a "**bacteriophage**." Despite the clear parallels with previously discovered viruses, however, the connection was not immediately made.

It is a matter of long-lived debate, continuing from the 1920s, whether d'Herelle or Twort should be credited with the discovery of bacteriophages. One thing is clear, though: d'Herelle immediately recognized the potential importance of bacteriophages and was not shy about promoting bacteriophages as antibacterial agents. The public was mesmerized by this possibility. *Arrowsmith*, a novel by Sinclair Lewis about a doctor who uses bacteriophages to successfully treat disease, was a best-seller in 1925. Lewis won the Pulitzer Prize for best novel a year later (he declined the award). Meanwhile, many more bacteriophages (or simply phages) were being discovered.

Drug manufacturers jumped on the phage therapy bandwagon and introduced virus cocktails for the treatment of various diseases. Bacteriophages were even added to water supplies on occasion. Moreover, early trials looked very promising. There were claims that *Staphylococcus* infections, dysentery, cholera, and typhoid were all successfully treated with phages (Fruciano and Bourne, 2007). Phage therapy was particularly embraced in the Soviet Union, and d'Herelle himself helped set up several production centers. Even today, phages are being grown for therapeutic use in Tbilisi, Georgia.

Then phage therapy fell out of favor. There were a number of reasons for this (Fruciano and Bourne, 2007). Many of the early, successful studies were poorly designed and controlled, leading to skepticism, especially since later studies frequently gave negative results. Another factor was the discovery of penicillin in 1928 and sulfa drugs shortly thereafter. These compounds were active against many kinds of bacteria (a property we now view as a disadvantage), whereas different phages were required for each bacterial target, making therapy more complicated. However, there were nonscientific reasons as well, leading many to think now that studies of phage therapy were prematurely terminated. As relations between the Soviet Union and the West soured in the 1930s, phage therapy became a symbol of ideological difference. The Soviets claimed that phage therapy was the future of medicine and made extravagant claims to support their view. In the West, antibiotic therapy was extolled with equal intensity.

Adding to the controversy was d'Herelle himself. While a gifted microbiologist, he managed to alienate much of the scientific community with his strongly held views and dismissive attitude toward other theories. For example, d'Herelle downplayed the independent role of the immune system in combatting disease, preferring instead to view recovery as always the triumph of bacteriophage over its bacterial host. His refusal to consider that Twort might have been the first to discover phages was also resented by many.

Recently, there has been renewed interest in using bacteriophages as a treatment for disease (Merril et al., 2003). In large part this interest stems from the rapid rise of antibiotic-resistant bacteria. There are now well-controlled studies showing that phages can rescue mice from normally lethal injections of staphylococci, enterococci, and *E. coli*. Phages have also been used successfully to treat calves with experimentally induced diarrhea due to *E. coli*. On the other hand, an experimental trial in Bangladesh where bacteriophages were used to treat severe diarrhea in children was unsuccessful (Sarker et al., 2016). Administering phages had no effect on the outcome of the disease.

While Felix d'Herelle might have overstated the importance of phages in our response to disease, he was certainly right in thinking that phages are an important part of the biosphere. There are thought to be approximately 10^{31} bacteriophages on the planet, and these are found in practically all environments. In seawater, there are about 10^6 to 10^7 bacteriophages per ml, and in humans, 10^8 to 10^9 phages per gram of feces (Kim et al., 2011). It is difficult to grasp the enormous number of phages in the world. In an attempt to better appreciate the magnitude of 10^{31}, a group of junior high school students made a number of interesting comparisons (http://www.pitt.edu/~gfh/printprotocol.pdf), among them:

- The number of phages = the number of words typed if you typed one *quadrillion* words per minute for 3.29 *quadrillion* years.
- The number of phages = distance in miles for 425 *billion* round trips to the Andromeda galaxy (which is 2.5 million light-years from Earth).

Bacteriophages do not all interact in the same way with their hosts, and some bacteria can show more than one kind of interaction. The most familiar interaction is called the **lytic cycle**. A phage particle attaches to its corresponding receptor molecule at the surface of the cell (Fig. 5-2, step 1). Its nucleic acid, which, depending on the phage, can be double-stranded DNA, single-stranded DNA, or even RNA, is subsequently taken up into the cytoplasm by different, poorly understood mechanisms (step 2). The phage structural proteins remain outside the cytoplasm. The nucleic acid then replicates, and its genes are transcribed and translated by the host machinery (step 3). Diversion to the synthesis of phages rather than cellular proteins is a burden on the cell but does not necessarily result in cell death. In step 4, phage proteins are assembled into new viral particles, and during this process the replicated nucleic acid is packaged into the phage particle. Finally, the cell is lysed from within by a phage protein and the progeny phages released (step 5). These phages can then infect other cells and undergo the same stages of development—thus the term "lytic cycle"—but it might be better to think of it as a chain reaction. Generally, an initial phage particle infecting a single cell results in the production of many progeny particles, each free to infect a new cell. When these new phage particles infect other cells, the number of phage particles resulting from these infections is again many times more than the number of infecting phages. Thus, there

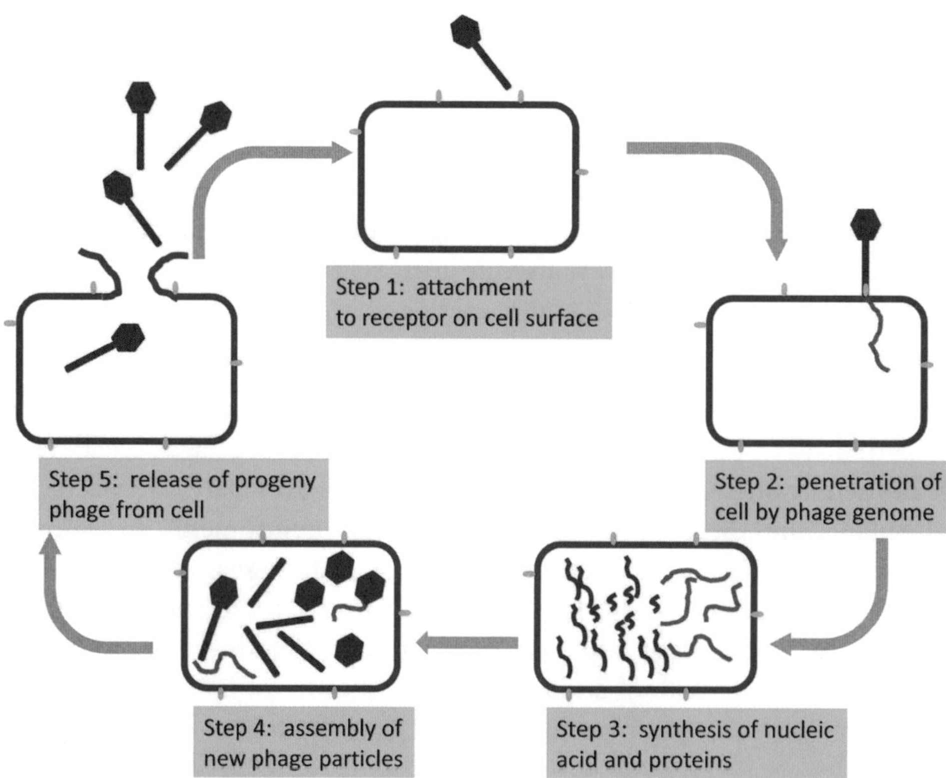

Figure 5-2 The general steps of the lytic cycle.

is a rapid increase in the number of viruses in a population of bacteria, and all the cells can quickly become infected and killed. When cells are growing in liquid culture, this can be observed by a decline in the turbidity of the culture due to lysis of the cells. However, there is another way to observe cell lysis. A small number of virus particles are mixed with a much larger number of growing cells (Fig. 5-3). The ratio of the number of phage particles to the number of cells is called the **multiplicity of infection (MOI)**, and in this case it is much less than 1 (MOI << 1). The infected culture is immediately spread on a plate containing agar medium. After incubation, there are too many bacteria to see individual colonies, but instead there is a **lawn of bacteria**. This lawn consists of closely packed cells on the surface of the medium. What about the phages that were added to the culture? Initially, only a very small number of bacteria are infected, because of the low MOI. In addition, the probability that a particular cell will be infected by more than one virus is near zero. When the rare infected cell produces progeny virus, these are released onto the lawn and infect neighboring cells. The new phages resulting from these infections are then able to infect adjacent cells in the lawn as well. Eventually, a clear spot in the lawn becomes visible to the naked eye. This spot is termed a "**plaque**," and it is clear because it consists of lysed neighboring cells. When d'Herelle observed "clear spots . . . speckling the cultures grown on agar," he was probably observing plaques.

The usefulness of this procedure is 2-fold. First, it allows us to determine the **titer**, the concentration of phages in a preparation, in much the same way that the num-

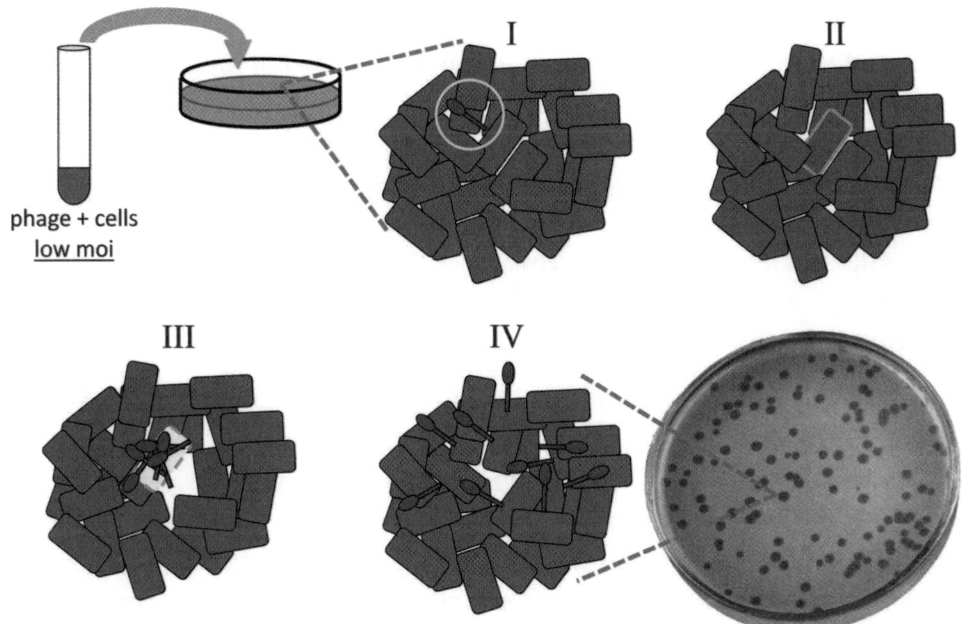

Figure 5-3 Stages in plaque formation. A rare bacteriophage particle infects a cell within a lawn of bacteria (I). After the phage nucleic acid has penetrated the cell, proteins and copies of the nucleic acid are produced (II). Progeny virus are assembled and the infected cell is lysed (III). The liberated phage particles then infect neighboring cells (IV). This process continues to create a visible plaque. Photograph from Stent, 1963.

ber of cells in a culture is determined (Challenge Three). Serial dilutions of the phage preparation are mixed with a fixed number of cells. Each mixture is then spread onto agar medium and incubated. If there are too many phages in the dilution, then the MOI will be too high and there will be confluent lysis, one large clear area on the plate. If the MOI is much less than 1, then there will be one or more plaques on the plate. Since each plaque arose from a single infecting phage, we can calculate back to determine the concentration of phages in the original preparation. Of course, if the MOI is too low, we might not see any plaques at all.

The second advantage of plaques is that each represents a pure clone of the single phage that initiated the plaque. Thus, it is used for the isolation of mutated phages and for the purification of phages from mixed populations, such as those found in natural sources.

Testing water purity

KEY POINTS

- The "coliforms" are a related group of microorganisms, including *E. coli*, that are used to monitor water quality.
- Modern methods allow the quick detection of *E. coli* and other coliforms in a water sample.
- Enumerating coliform phages in a water sample offers a safer alternative to growing these organisms.

E. coli was identified in 1884, and it was soon realized that its primary habitat was the gut of mammals, including man. The Austrian microbiologist Franz Schardinger proposed that testing for the presence of *E. coli* in a water sample would indicate fecal contamination. This was an attractive suggestion because, unlike many pathogenic microorganisms, *E. coli* was hardy and easy to grow. A facultative anaerobe (Challenge One), it was able to ferment lactose, a trait which at the time was considered the hallmark of the species. This made *E. coli* easy to identify. In effect, then, **commensal** strains of *E. coli* (strains that colonize a host but do not cause disease) were proposed to be surrogates for pathogenic organisms in the gut.

In the following years, organisms similar to *E. coli* were identified. These were also Gram-negative, facultative anaerobes able to ferment lactose, and at first they were mistakenly identified as *E. coli*. This group of organisms were named "**coliforms**," not a true taxonomic classification but a convenient one, nonetheless. All are members of the family *Enterobacteriaceae*, a large group of bacteria that includes non-coliforms as well. By 1943 in the United States, standards of water purity were based on the total coliform count in samples. This would not have been a problem if all coliforms came from the gut, but some species of coliforms occupy other environmental niches, such as plants and plant debris. As a result, the total coliform count can lead to an overestimation of fecal contamination in many instances, although it is sometimes still used as an indicator of water purity.

Modern methods of assaying water purity frequently involve chromogenic and/or fluorogenic enzymatic reactions (reactions that result in a colored or fluorescent product) to identify both total coliforms and *E. coli*. One version is shown in Fig. 5-4. The water sample is added to medium containing bile salts and lactose and two indicator compounds, *o*-nitrophenyl-β-D-galactopyranoside (ONPG)

Figure 5-4 A modern procedure for assaying total coliforms and *E. coli* in a water sample.

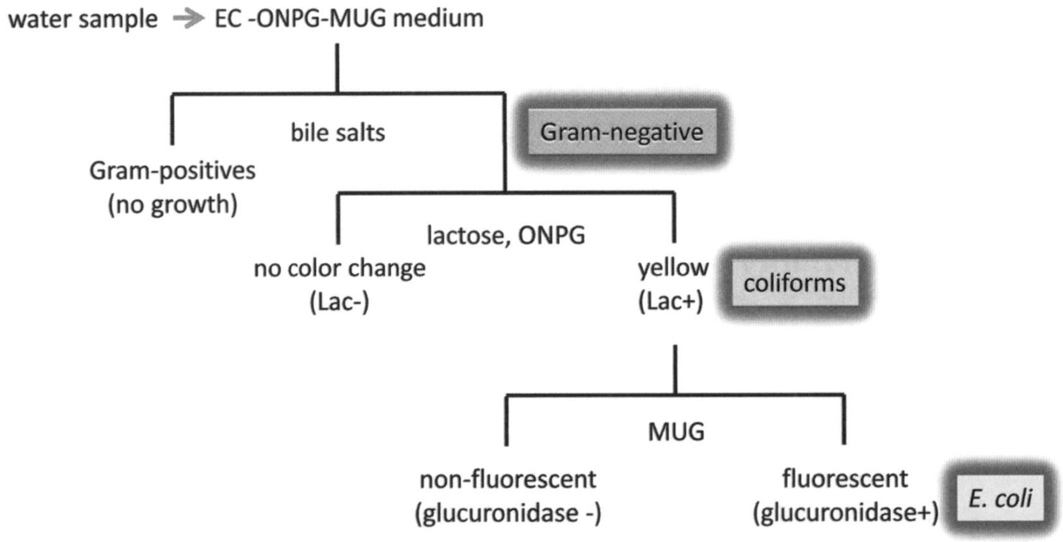

water sample → EC -ONPG-MUG medium

bile salts

Gram-positives (no growth)

Gram-negative

lactose, ONPG

no color change (Lac-)

yellow (Lac+)

coliforms

MUG

non-fluorescent (glucuronidase -)

fluorescent (glucuronidase+)

E. coli

and 4-methylumbelliferyl-β-D-glucuronide (MUG). The medium is selective for Gram-negative organisms: Gram-positive bacteria will not grow in the presence of bile salts (Challenge One), so all the organisms that grow are Gram negative. Lactose-fermenting organisms are detected because they contain the enzyme β-galactosidase, which converts the ONPG to a compound that turns the medium yellow. If a color change is observed, then lactose-fermenting, Gram-negative cells are present and these are almost certainly coliforms. Finally, MUG in the medium is converted by the enzyme β-glucuronidase to a product that is fluorescent under UV illumination. *E. coli* is the only coliform having this enzyme, so fluorescence will be observed only when this species is present. One advantage of this assay is that multiple samples can be tested at one time, by using trays that contain an array of wells. The test is not error-proof, however: the dangerous pathogen *E. coli* O157:H7 lacks β-glucuronidase and would not be detected by fluorescence.

Detecting water contamination by directly culturing bacteria from a sample has a number of disadvantages. First, if a pathogen is suspected, then its cultivation involves health risks and must be done under containment conditions that are not always available (see Introduction). The circumstances outlined in this challenge are an example of this problem. Second, quantifying the results frequently involves making serial dilutions, which would also be hazardous with heavily contaminated samples. Finally, the time between sampling and assaying is important: bacteria can die, of course, but also some coliforms can multiply, even in water.

Another approach is to determine the load of bacteriophages in water samples (Jofre et al., 2016). Samples are filtered to remove bacteria and sediment, and the phage concentration in the filtrate then determined by one of several methods, including the double agar overlay method described in the next section. **Coliphages**, the bacteriophages that infect *E. coli*, are usually assayed. The assumption here is that the more coliforms in a sample, the greater the number of its attendant phages, including those specific for *E. coli*. This assumption was tested by Wentzel et al. (1982), and a good correlation was found between the number of coliphages in a sample and the total number of coliform bacteria, which was determined by direct cultivation. The advantage of this method is that levels of water contamination can be determined without growing potentially hazardous bacteria. Water can be filtered at collection sites and then the whole filtration device bagged and sealed for subsequent safe disposal. Coliphages are not only nonhazardous but also stable, so they can be transported from a sample site to a lab with little decrease in the number of infectious particles.

A concern with using coliphages is that if *E. coli* can grow in aqueous environments outside of the gut, as some have claimed, then bacteriophages might continue to infect cells and multiply, leading to an overestimate of contamination. However, a recent critical analysis of this possibility led to the conclusion that virtually all the bacteriophages in a sample are from the gut and contributions from other sources are minimal (Jofre et al., 2016).

A greater difficulty is selecting the *E. coli* strain that will be used in the assay. Different *E. coli* strains are not equally sensitive to natural populations of coliphages. One problem is that some strains are resistant to infection because they lack the attachment site for the phages. Another is that many have restriction systems, enzyme complexes that recognize DNA from other sources as foreign and destroy it. *E. coli* strain C, however, does not contain these restriction systems and is sensitive to many different coliphages. The Environmental Protection Agency (EPA) of the U.S. government has published two approved protocols for determining water quality by assaying total coliphages with *E. coli* strain C.

Lab One

1. ***Pass the samples through a filter to purify the phages.***
2. ***Titer the flowthrough to determine the concentration of phages in each sample.***
3. *Determine from the titer at each location which farm(s) are the major source of contamination.*

Learning outcomes

After this lab, students will be able to:

- **a.** Set up a soft agar overlay and spot-titer plate.
- **b.** Calculate a phage titer.

I. Determine the load of bacteriophage at each collection site

Water samples (1 ml) were taken at the 10 different sites shown in Fig. 5-1.

PROCEDURE

Filter the water samples to remove all the bacteria

1. Working individually or in pairs, first remove the plunger from a 5- or 10-ml syringe. Fit a disposable 0.2-µm-pore-size filter onto the syringe body. Set the assembly on the top of a culture tube (Fig. 5-5).

   Bacterial cells and particles of debris are too large to pass through the filter.

2. Carefully pour one of the water samples into the syringe.

   Liquid will not pass through the filter until force is applied with the plunger.

3. Replace the plunger and apply pressure to force the liquid through the filter. Use one hand to push on the plunger and the other to steady the syringe. Once the filtrate has been collected, immediately cap the tube and discard the filter and syringe as directed without disassembling.

   The bacteriophages will be in the collected filtrate (the volume will be less than 1 ml).

PROCEDURE

Titer the phages in the sterile filtrates

As described above, determining a phage titer is very much like doing serial dilutions and colony counts (Challenge Three). Different dilutions of the phage preparation are mixed with a large number of cells, which are then spread on a plate. When the MOI is <<1, the number of plaques reflects the number of phage particles at that dilution. It is important that the lawn of cells is uniform for plaques to be visualized easily. Simply using a spreader does not accomplish this; invariably the lawn will vary in cell density, with some locations having more cells than others. To create a uniform lawn, cells and phages are mixed with medium containing a low concentration of melted agar ("soft agar") and immediately poured onto solidified agar medium in a plate. The plate medium contains agar at the normal concentration, about 1.4%. The concentration of agar in the soft agar overlay is generally between 0.4 and 0.8%: low enough so that it will pour easily but high enough so that it will still solidify after cooling. After incubation, the result is a lawn with distinct plaques (Fig. 5-3).

You will do a "spot titer," a variation of the soft agar overlay method. This procedure saves both time and plates and is convenient when the titer does not need to be accurate. With this method, the soft agar overlay is created first, using uninfected cells, and allowed to solidify. A fixed volume of different phage dilutions is then applied to sections of the lawn. An example is shown in Fig. 5-6.

1. Mark the back of two tryptic soy agar (TSA) plates so that each is divided into three sectors of approximately equal size. Label the sectors of one plate 10^{-2}, 10^{-3}, and 10^{-4} and the other 10^{-5} and 10^{-6} (one sector can remain blank, or use it for sterile dilution buffer as a control). *Make sure that there are no water droplets on the surface of the agar and that the plates are at room temperature.*
2. Pipette 50 µl of an overnight culture of *E. coli* W3110 into two small, sterile culture tubes (provided).
3. The soft agar (4 ml) will be in small culture tubes that have been placed in a water bath or heat block set to 44 to 46°C. Move your plates and the tubes containing the cells close to the soft agar tubes, since everything will need to be done quickly. A vortexer should be near the soft agar tubes as well.
4. Pour 4 ml soft agar into one of the tubes containing your cells, vortex for 1 to 2 seconds, and then pour onto the surface of a TSA plate. As you pour, you can tilt the plate slightly to distribute the soft agar, but stop moving the plate as soon as the surface has been covered (it is not a problem if a *small* part of the plate is lacking the overlay; just avoid it when spotting). Using a second TSA plate, repeat this procedure with your other sample of cells.

   The plate will start to set in a few seconds, and after 5 minutes, you can gently take it back to your bench. The overlay should appear smooth: if there are wrinkles or lumps, or if the surface appears granular, the agar cooled too quickly (probably you did not pour the overlay quickly enough). In that case, try again with a new plate.
5. Prepare 10^{-2}, 10^{-3}, 10^{-4}, 10^{-5}, and 10^{-6} dilutions of your filtered sample, using the phage dilution buffer provided. Follow the dilution protocol shown in Fig. 5-7. To save time and materials, prepare the 10^{-2} dilution by adding 10 µl of your sample to 0.99 ml dilution buffer. If the dilution volume is difficult to measure, simply use 1 ml of the buffer: the error introduced is negligible.

Procedure continues on next page

Titer the phages in the sterile filtrates (*continued*)

6. Deposit 25 μl of each dilution onto the appropriately labeled sector of the soft agar overlay. Use a new micropipette tip for each dilution and position the tip so that the spots on each sector will be as far away from each other as possible.

   Hold the tip just above the surface of the agar medium and slowly expel the liquid. If the tip penetrates the medium, the opening of the micropipette tip will first be blocked and then as the tip is moved the liquid will come out too quickly. If the tip is more than a few millimeters above the surface, the liquid can splatter, with droplets "dancing" on the surface, resulting in several spots on the plate. If the spots are well separated, you can rotate the plate gently to increase their surface area on the plate.

7. With the plate cover slightly ajar, let the spots dry until there is little or no visible liquid on the surface.

   Be patient. If you move the plate into the incubator immediately after spotting, the different dilutions will likely come into contact and you will not get usable data.

8. Place the overlay plates cover side up in a 37°C incubator and examine the results the next day.

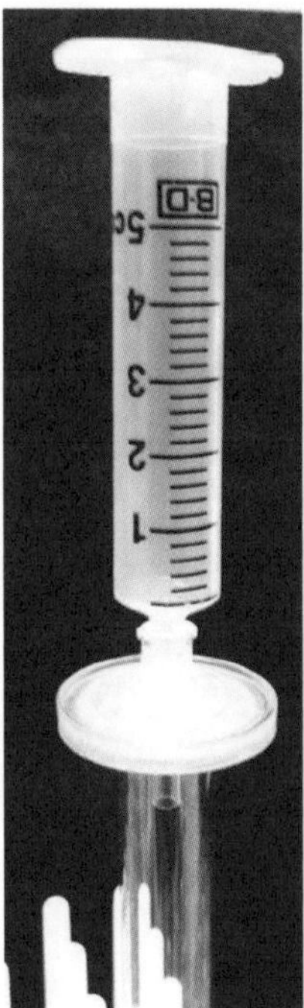

Figure 5-5 Disposable filter and syringe body.

Figure 5-6 Spot titer result for different dilutions of phage lambda. Twenty-five microliters of the indicated dilution were applied to each quadrant.

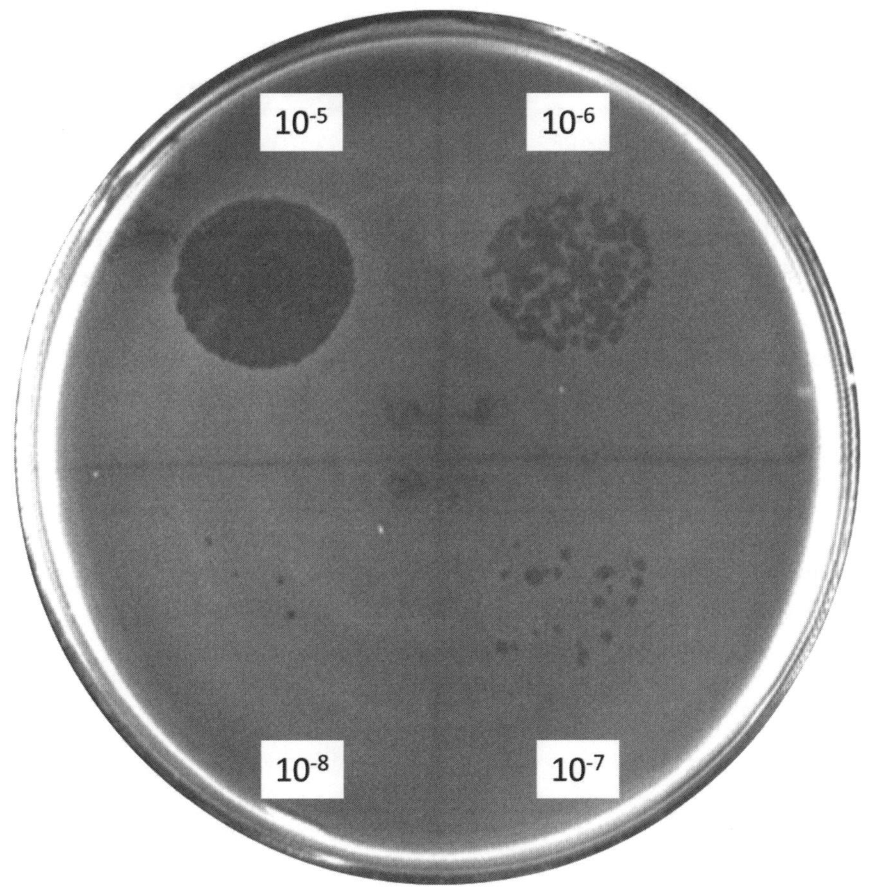

Figure 5-7 Diluting the filtered water sample for spotting on a bacterial lawn. The first dilution step is 1:100.

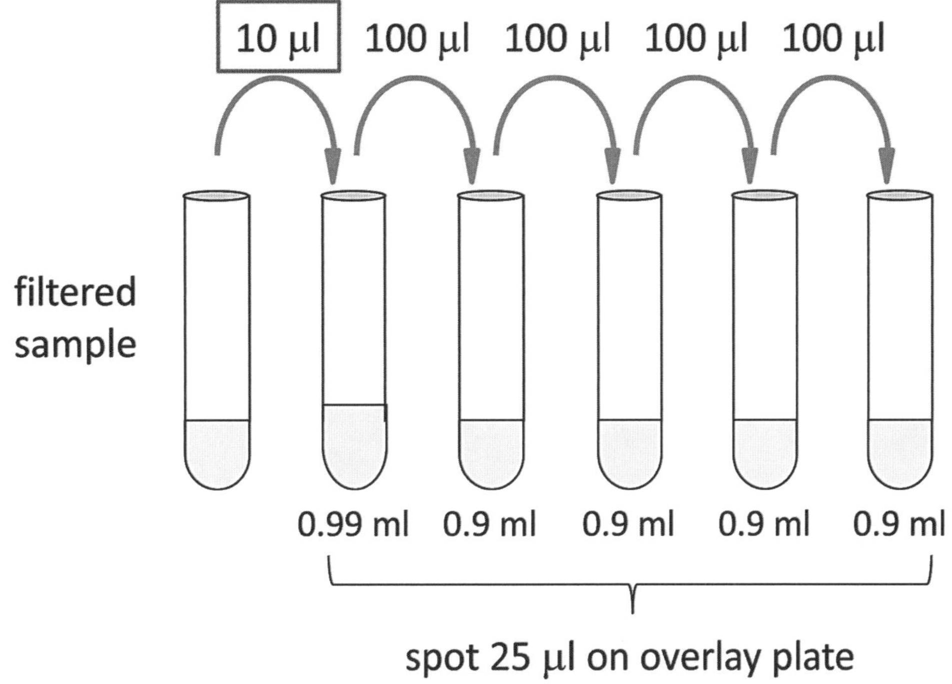

Solving Challenge Five

1. *Pass the samples through a filter to purify the phages.*
2. *Titer the flowthrough to determine the concentration of phages in each sample.*
3. ***Determine from the titer at each location which farm(s) are a major source of contamination.***

Enter the results of your phage titer in Table 5-1 (or submit your results to the instructor, who will then complete and post the table). Use "Cf" (confluent) when there is a single clear spot rather than individual plaques, "TNTC" when plaques are visible but are numerous and overlapping, and "NP" for sectors having no plaques (Table 5-1). Otherwise, record the number of plaques in the sector and estimate the phage titer in the water sample you analyzed. This titer (phages per milliliter) can be calculated by:

$40 \times$ (no. plaques in spot) $\times$ cumulative phage dilution (10^{-2}, 10^{-3}, 10^{-4}, 10^{-5}, or 10^{-6})

EXAMPLE

You obtained 22 plaques when the 10^{-4} dilution was spotted on the cell lawn. You spotted 25 µl, so the number (n) of plaques from 1 ml (1,000 µl) would be:

$$\frac{22 \text{ plaques}}{25\,\mu\text{l}} = \frac{n \text{ plaques}}{1{,}000\,\mu\text{l}}$$

$$n = \frac{(1{,}000)(22)}{(25)} = 40 \times 22 = 880 \text{ plaques}$$

Since each plaque represents one virus particle, the concentration of phages in the 10^{-4} dilution must be 880 phages per ml. Applying the multiplication factor indicated by the serial dilutions, the original sample contained:

$$880 \times 10^{4} = 8.8 \times 10^{6} \text{ phages per ml}$$

With the data from the class, you should then be able to identify the location(s) where the water is being contaminated. Assume that the harmful bacteria (and therefore the phages) will become progressively diluted downstream from the point of entry, provided there are no additional sources of contamination. If there are any inconsistencies in the results for each sample site, explain which values you believe are most likely correct.

Table 5-1 Results of phage titers

Sample:	Group:	Spot titer results (plaques):					Estimated phage titer of sample (phage/ml)
		10^{-2}	10^{-3}	10^{-4}	10^{-5}	10^{-6}	
1							
2							
3							
4							
5							
6							
7							
8							
9							
10							

QUESTIONS

Questions are designated B1 to B6 according to the six levels of Bloom's taxonomy.

1. In enumerating the number of virus particles, why is it important that the MOI is <<1? (B1)
2. During the procedure above, it was stated that you could dilute your phage sample in 1 ml rather than 0.99 ml dilution buffer. Why is this difference in volume negligible? (B2)
3. When estimating a titer from one of the spots, why is "40" used in the calculation? (B2)
4. *Klebsiella* species are important members of the coliforms, yet they are distributed widely in nature and are found in soil and on plants. How might this complicate the coliform analysis of an outdoor water well located in a rural area? (B2)
5. In the example of the spot-titer plate (Fig. 5-6), there were four plaques at the 10^{-8} dilution. What was the approximate concentration of virus (particles/milliliter) in the lysate? (B3)

BIBLIOGRAPHY

Duckworth DH. 1976. Who discovered bacteriophage? *Bacteriol Rev* **40:**739–802. http:// http://mmbr.asm.org/content/40/4/793.long.

Fruciano DE, Bourne S. 2007. Phage as an antimicrobial agent: d'Herelle's heretical theories and their role in the decline of phage prophylaxis in the West. *Can J Infect Dis Med Microbiol* **18:**19–26.

Jofre J, Lucena F, Blanch AR, Muniesa M. 2016. Coliphages as model organisms in the characterization and management of water resources. *Water* **8:**199. http://www.mdpi.com/2073-4441/8/5/199/htm.

Kim MS, Park EJ, Roh SW, Bae JW. 2011. Diversity and abundance of single-stranded DNA viruses in human feces. *Appl Environ Microbiol* **77:**8062–8070. http://aem.asm.org/cgi/doi/10.1128/AEM.06331-11.

Lustig A, Levine AJ. 1992. One hundred years of virology. *J Virol* **66:**4629–4631. http://jvi.asm.org/content/66/8/4629.long.

Merril, CR, Scholl D, Adhya SL. 2003. The prospect for bacteriophage therapy in Western medicine. *Nat Rev Drug Discov* **2:**489–497. http://dx.doi.org/.

Sarker SA, Sultana S, Reuteler G, Moine D, Descombes P, Charton F, Bourdin G, McCallin S, Ngom-Bru C, Neville T, Akter M, Huq S, Qadri F, Talukdar K, Kassam M, Delley M, Loiseau C, Deng Y, El Aidy S, Berger B, Brüssow H. 2016. Oral phage therapy of acute bacterial diarrhea with two coliphage preparations: a randomized trial in children from Bangladesh. *EBioMedicine* **4:**124–137. http://www.ncbi.nlm.nih.gov/pubmed/26981577.

Wentsel RS, O'Neill PE, Kitchens JF. 1982. Evaluation of coliphage detection as a rapid indicator of water quality. *Appl Environ Microbiol* **43:**430–434.

Evaluating the pathogenic potential of bacteria causing urinary infections

Nurses in a postsurgical unit noticed an increase in urinary tract infections in postoperative patients. Urine cultures revealed two species of bacteria: *Escherichia coli* and an uncharacterized species of *Pseudomonas*. These strains have been found on doorknobs in the hospital as well as on meal trays and catheter stands used to hold urinary catheters. Normal cleaning procedures have reduced but failed to eliminate the bacteria. You suspect the bacteria are forming biofilms resistant to routine cleaning. You will determine whether they do indeed form biofilms and also recommend a solution for decontamination. You will also test whether several compounds found in urine are chemoattractants for these strains.

QUESTIONS BEFORE YOU BEGIN THE CHALLENGE

1. State the hypotheses that you will test in this challenge.
2. Define "chemoattractant." What is required of cells that exhibit chemotaxis? Review the class results from Challenge One. Do *E. coli* and *Pseudomonas* meet any of these requirements?

Strategy for Challenge Six

1. Visualize biofilm formation using crystal violet.
2. Quantitatively analyze biofilm formation using spectrophotometry.
3. Determine if the strains produce quorum-sensing compounds.
4. Determine whether compounds in urine are chemoattractants for these strains.
5. Analyze the effectiveness of chemical cleaners by the disk diffusion assay.

Lab One

BACKGROUND

Quorum sensing

KEY POINTS

- Some genes, including many involved in virulence and the formation of biofilms, are beneficial only when there is a high density of cells. The expression of these genes is regulated by a mechanism called quorum sensing.
- Quorum sensing was first described in the light-producing microorganism *Vibrio fischeri*. It is now known that many bacteria possess quorum-sensing systems.
- Bacteria use a variety of small molecules to determine and signal population density.

Although bacteria are single-celled organisms, some genes are expressed only when there is a high concentration of cells. This allows bacteria to exhibit traits that are only beneficial at a high population density. How does a single-celled organism know how many members are in the population? Many bacteria use a process termed **quorum sensing** to assess cell numbers (reviewed in Waters and Bassler, 2005). These microbes produce small signaling molecules called **autoinducers**: Gram-negative organisms use **acyl homoserine lactones (AHLs)**, while Gram-positive organisms use small peptides. Each organism produces its own communication molecule to speak a unique "language."

Quorum sensing was first described in *V. fischeri*, a Gram-negative marine bacterium that produces light using the LuxI/LuxR system. *V. fischeri* has a fascinating relationship with the bobtail squid, a small, nocturnal animal found in shallow marine waters (Fig. 6-1). Bright moonlight produces a shadow of the squid on the seafloor, making it an easy target for predators. The squid has a light organ, a small pouch on its ventral surface that is colonized by *V. fischeri* shortly after it hatches. Each morning, the bobtail squid expels most of the *V. fischeri* cells from the light organ, retaining a small population that grows throughout the day. Initially, the low amount of light emitted by the remaining cells would be ineffective and the system for light production is therefore turned off. However, once the number of cells reaches a critical level, when enough light would be produced, the system is then turned on. The bacteria detect the critical level of cells by a quorum-sensing circuit. How does this circuit work (Fig. 6-2)?

1. The *luxI* gene of *V. fischeri* encodes an AHL, which diffuses in and out of the cells. When population numbers are low, the AHL molecules proceed along

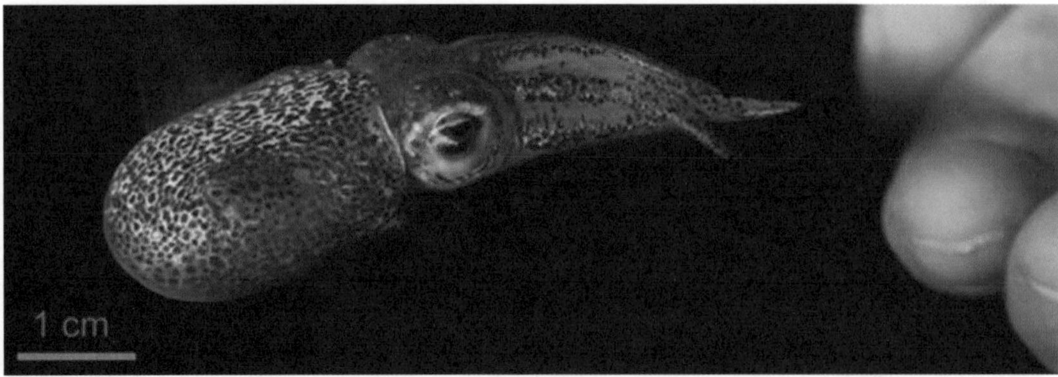

Figure 6-1 The Hawaiian bobtail squid. From PLoS Biology, volume 12, 2014. Credit Margaret Mcfall-Ngai. CC-BY-4.0.

a concentration gradient into the medium outside the cell. As the population grows, so too does the concentration of AHL within the light organ, resulting in more AHL molecules diffusing back into the cell.

2. Once intracellular concentrations of AHL have reached a threshold, AHL molecules are bound by LuxR proteins in the cell.
3. The AHL/LuxR complex binds to the *luxI* promoter, increasing expression of this gene as well as the genes for light production.

Figure 6-2 The LuxR/LuxI quorum sensing system. Autoinducer diffuses out of the cell and accumulates as the population grows. When the autoinducer reaches a threshold concentration, the LuxR/autoinducer complex induces transcription of the *lux* operon and light production commences.

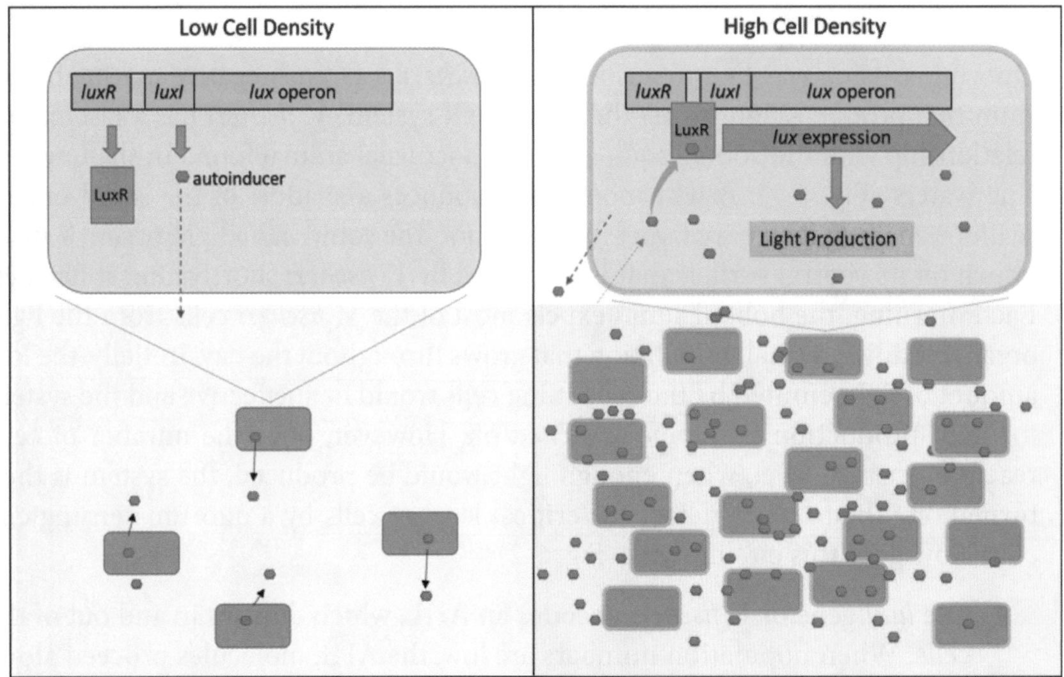

4. The increased expression of *luxI* results in a sudden increase in the amount of AHL, further increasing light production by binding to additional LuxR molecules.
5. As long as there is a high concentration of cells, enough AHL is produced for continued expression of the genes for light production.
6. If the concentration of cells significantly decreases, not enough AHL is produced to keep the *luxI* promoter activated by the AHL/LuxR complex.
7. As a result, the genes for light production are no longer active. Expression of *luxI* also drops, reinforcing the "off" mode by causing a further decrease in concentration of AHL.

Light production peaks during nighttime, when the concentration of *V. fischeri* in the light organ is highest. The light shines through this organ down onto the seafloor to eliminate the shadow of the squid. As the sun rises, the squid expels the majority of the *V. fischeri*, resetting the process. This is a mutualistic relationship, in which both species benefit from the symbiosis: the bacteria grow in the light organ with exclusive access to nutrients provided by the squid while the squid evades predators.

Why not simply maintain a constant, dense population of *V. fischeri*? Expulsion of the bacteria each morning prevents overcrowding of the light organ and reduces the amount of nutrients the squid must provide. It also avoids carrying stationary-phase cells, which have lower light production.

Light production is an interesting and classic example of quorum sensing, but this process is used to control many other group behaviors of bacteria, including the expression of virulence factors and the formation of biofilms, both important in pathogenesis. Quorum sensing is fairly common in Gram-negative bacteria: a search in 2007 revealed that of the 265 sequenced proteobacterial genomes, 68 had homologs to LuxI and LuxR (Case et al., 2008). (Homologs are structurally similar proteins with a shared evolutionary history and frequently similar functions.) *Pseudomonas aeruginosa* is an opportunistic pathogen that is a significant threat to immunocompromised people, individuals with cystic fibrosis, and patients with large open wounds or burns. *P. aeruginosa* possesses four quorum-sensing circuits, two of which comprise LuxI/LuxR homologs. These four systems control many virulence properties, including the production of proteases and elastases that contribute to the tissue destruction associated with these infections, and also play a role in biofilm formation (below).

Biofilms

KEY POINTS

- Bacteria often attach to surfaces and grow within a structure called a biofilm. Biofilms can be found throughout nature and in the human body, where they can be beneficial or associated with disease.
- Biofilm development occurs in five stages. Bacteria growing in a biofilm exhibit different traits than those growing in liquid culture.
- Biofilms are of significant concern in health care settings because cells within in a biofilm are more resistant to antibiotics and often produce virulence factors.

Liquid medium is often used in experiments because it gives reproducible results when the cells are in balanced growth (see Challenge Three). However, it is important to realize that in nature most bacteria rarely grow suspended in liquid (planktonic growth); rather, they are found attached to surfaces in a structure called a **biofilm**. Biofilms were first documented in the literature in the 1930s, but the term "biofilm" was not used until the 1970s. Biofilms are found almost everywhere, from rocks along a riverbank to the surface of our teeth. A biofilm develops in five stages: (i) initial attachment, (ii) irreversible attachment, (iii) maturation-1, (iv) maturation-2, and (v) release (Fig. 6-3). During the initial attachment stage, cells loosely attach to a surface via weak forces such as van der Waals forces or hydrophobic interactions between the bacterial cell and components of the object's surface. Cells can be easily washed from the surface at this time. Stronger bonds are formed as the cells transition into stage 2, the irreversible attachment step. These bonds are due to **adhesins**, proteins that recognize and attach to a specific component of the attachment site, or to less specific extracellular polymeric substances (EPSs), which are usually composed of polysaccharides and polypeptides. EPSs form an extracellular matrix that holds the cells together and firmly anchors them to the surface. Removal of the cells from the surface becomes difficult at this point.

Figure 6-3 The stages of biofilm development. 1) Initial attachment, 2) Irreversible attachment, 3) maturation-1, 4) maturation-2, 5) release. Figure adapted from PLoS Biology, volume 5, 2007. Credit: Don Monroe. CC-BY 3.0.

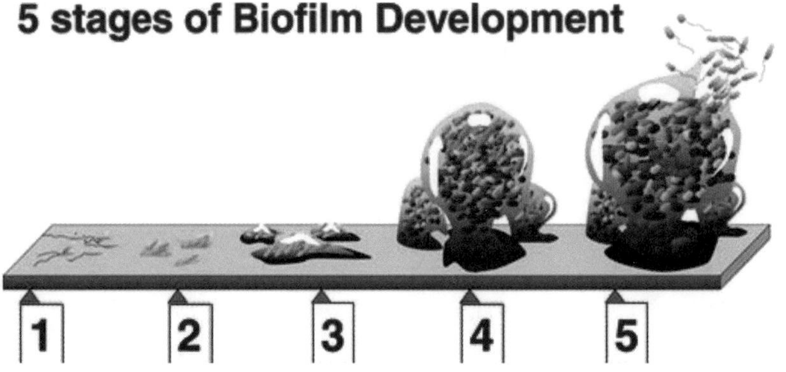

As the population continues to grow, the biofilm enters the maturation-1 phase, marked by EPS-encased microcolonies that range in size from approximately 10 to 100 μm in diameter. The biofilm further matures during maturation-2 to form large, very complex structures containing millions of cells. As you can imagine, the cells in the middle of the biofilm encounter a very different environment than cells on the outer edge of the biofilm. Finally, in the release stage, some cells disperse from the mature biofilm, a process that may aid in population survival as these cells are then free to colonize new sites.

Biofilms occur naturally all over the human body and can contribute to disease. If you've ever scraped your tooth with your fingernail to find white, slimy material, you have seen a biofilm (Fig. 6-4). Oral biofilms are complex, consisting of many different bacterial species. First, initial colonizers attach to the tooth surface by

Figure 6-4 Scanning electron micrograph of *Streptococcus mutans* and *Candida albicans* (a eukaryote) biofilm on a tooth surface. A) 100X magnification. The red box indicates the area further magnified in panel B. B) 1500X magnification. The red box indicates the area further magnified for photo C. C) 5000X magnification. The blue arrows indicate cells of *S. mutans*. Also visible are the much larger cells of *C. albicans* (red arrows) and water channels within the biofilm (white arrows). From PLoS Pathogens, volume 9, 2013. Credit: Metwalli et.al. CC-BY 3.0.

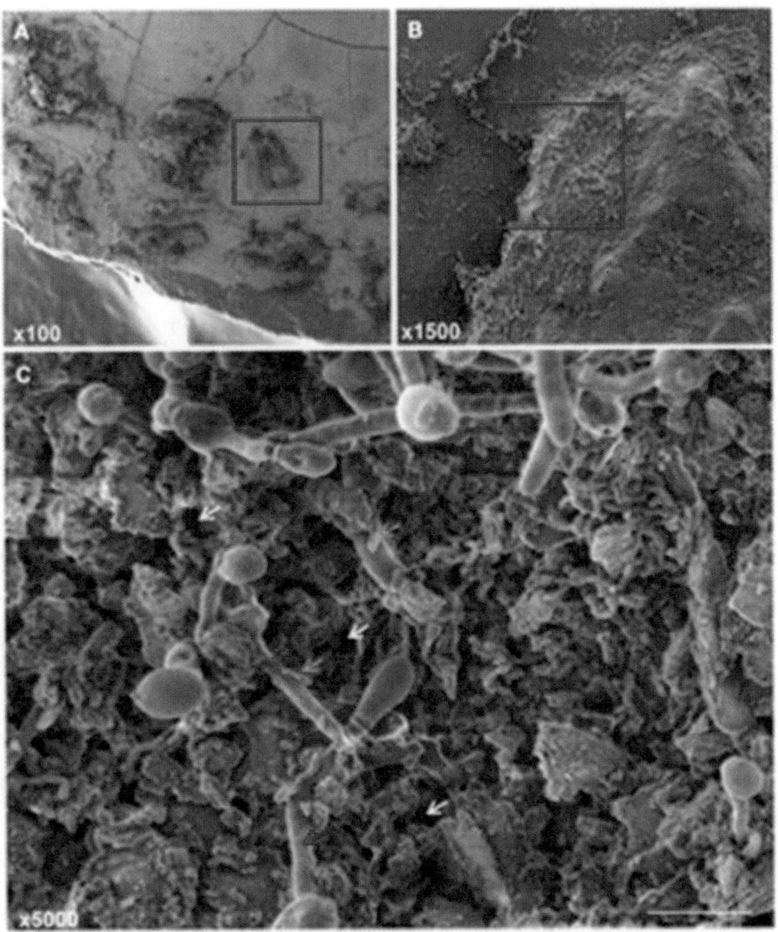

means of specific adhesins that recognize the salivary proteins coating the teeth. Then late colonizers, cells of different (and frequently pathogenic) bacterial species, attach to the biofilm. If not removed by flossing, brushing, and regular professional cleaning, uncontrolled biofilm growth in the oral cavity can result in cavities and gingivitis, an inflammation of the gums. Gingivitis can lead to periodontitis, where not only the gums but other tissues surrounding the teeth, such as bone, become inflamed as well.

Many gut bacteria form a biofilm on the mucus layer that lines the intestines. By taking up space and consuming nutrients, these bacteria inhibit colonization by pathogens. For example, the pathogen *Clostridium difficile* is often the cause of antibiotic-associated diarrhea (AAD), a sometimes deadly infection that is responsible for an estimated half a million hospital-acquired infections annually, according to the Centers for Disease Control and Prevention. *C. difficile* forms hardy spores that are resistant to antibiotics as well as to alcohol-based hand sanitizers and other commonly used disinfectants. When a patient is treated with antibiotics in the hospital, the normal biofilm in the intestines is disrupted. This allows the germination of the antibiotic-resistant *C. difficile* spores and subsequent colonization and infection of the intestinal tract. Because of this, many antibiotics now contain a warning about AAD. *C. difficile* infections have a high rate of recurrence when patients have been treated for AAD with antibiotics, because the normal intestinal biofilm remains disrupted by the antibiotic treatment, resulting in germination of remaining or newly-acquired spores. To reestablish a healthy population of gut microbes, some physicians have started using fecal transplants. Fecal matter from a healthy donor is collected, processed, and deposited into the patient's colon.

Biofilms are a significant problem in the health care industry. Bacteria can form biofilms on almost any surface, including prosthetic joints and catheters (Fig. 6-5). Bacterial growth on these devices is of particular concern because biofilm cells differ from planktonic cells in several very important ways. Cells growing in biofilms

Figure 6-5 Scanning electron micrograph of a *P. aeruginosa* biofilm on a urinary catheter. From *Antimicrobial Agents and Chemotherapy,* 2010 Vol 54(1) 397–404.

exhibit higher levels of antibiotic resistance. Up to 1,000 times more antibiotic is required to kill cells growing in a biofilm, an amount that often cannot be achieved during treatment. Additionally, biofilm cells are more likely to transfer genes by conjugation or transformation. Some of these horizontally acquired genes encode resistance to antibiotics. Finally, the synthesis of virulence factors, regulated by quorum sensing, is often initiated within the biofilm.

Lab One

1. ***Visualize biofilm formation using crystal violet.***
2. ***Quantitatively analyze biofilm formation using spectrophotometry.***
3. ***Determine if the strains produce quorum-sensing compounds.***
4. *Determine whether compounds in urine are chemoattractants for these strains.*
5. *Analyze the effectiveness of chemical cleaners by the disk diffusion assay.*

Learning outcomes

After this lab, students will be able to:

- **a.** Grow a biofilm.
- **b.** Use crystal violet to quantitatively measure biofilm formation.

I. Determine if the hospital isolates form biofilms

You will determine if one of the hospital isolates forms a biofilm after overnight incubation in a 96-well plate. Planktonic cells will be removed, leaving behind any biofilm, which will be stained with crystal violet. This positively charged dye binds to negatively charged molecules, including the surface of bacterial cells, DNA, and components of the extracellular matrix. The amount of crystal violet retained, and therefore the color intensity, is proportional to the degree of biofilm formation (Fig. 6-6).

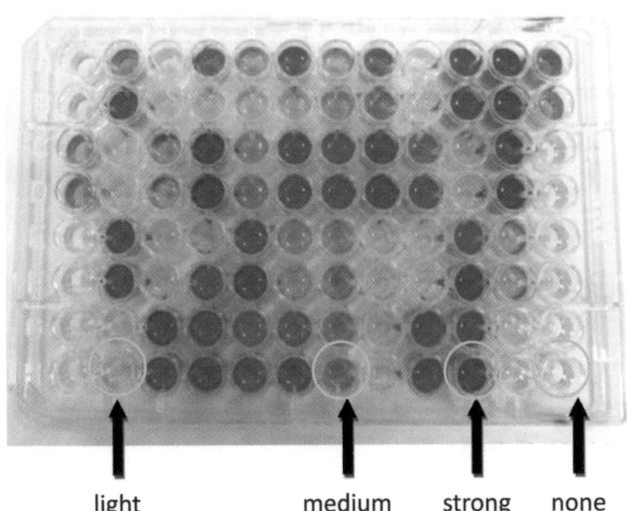

Figure 6-6 Different levels of biofilm formation, visualized by crystal violet staining, in the wells of a 96-well plate. From The Scientific World Journal, 2013. I.D. 378492. Credit: Darwish and Asfour. CC-BY 3.0.

PROCEDURE

Staining biofilms with crystal violet

Day before lab

Working in groups of two, obtain a sterile, 96-well covered tray similar to the one shown in Fig. 6-6. Approximately 18 to 24 hours before lab begins, pipette 200 µl sterile tryptic soy broth (TSB) into wells 1, 2, 4, 5, 7, and 8 of row A (Fig. 6-7). Using an overnight culture (provided), add 2 µl of your assigned hospital isolate to wells 1 and 2. Similarly inoculate wells 4 and 5 with the overnight culture of *E. coli* (the positive control) and wells 7 and 8 with *Bacillus cereus* (the negative control). During the lab, wells A1, A4, and A7 will be used for the first part of the experiment and wells A2, A5, and A8 for the second part. Cover the plate and place in a 30°C incubator until lab the next day.

Lab day

1. Retrieve your 96-well plate and resuspend any cells that have settled in the wells by *gently* drawing and expelling the liquid several times with a micropipette.
2. Remove the entire culture (200 µl) from each well and transfer it to the *same position* in row B (Fig. 6-7). The transferred cultures will also be controls, since the planktonic cells would not have had a chance to form biofilms in the new wells.
3. Examine the cells under the microscope.
   a. Divide a slide in half with a wax marker and label one side A1 and the other B1. Similarly label a second slide A4 and B4 and a third slide A7 and B7.
   b. Add a drop of water to sides A1, A4, and A7. Using pipette tips, separately scrape the sides of wells A1, A4, and A7 and add any transferred material to the correspondingly labeled droplet. Using a loop, add a small amount of culture from wells B1, B4, and B7 to side B.
   c. Examine the cells using phase-contrast or bright-field microscopy. Is there any difference in the distribution of cells within the drop for samples A (cells from the biofilm, if one was formed) and B (planktonic cells)?
4. Discard the planktonic cultures (row B) by inverting the tray over the provided container. Immerse the tray in a container of water to rinse the wells and then carefully shake the water out. Repeat this two more times.
5. Add 250 µl of 0.1% crystal violet to the wells and keep the plate on your bench for 10 minutes.
6. Immerse the tray in a container of fresh water to rinse the wells and then carefully shake the water out. Repeat this two more times.

The cells and extracellular matrix present in a biofilm will retain the crystal violet dye. Examine the wells in row A. Is there a difference in the appearance of the wells for the positive (*E. coli*) and negative (*B. cereus*) controls? Compare these wells to the ones containing the hospital strain you are testing. Does the hospital strain form a biofilm? Record your observations in your notebook.

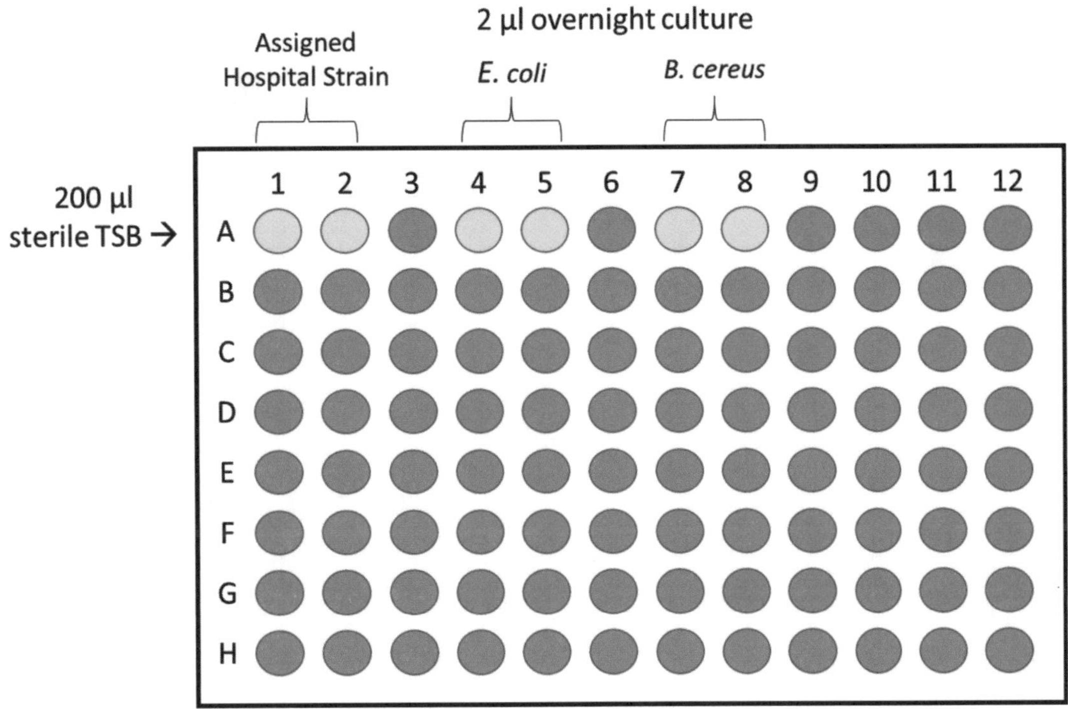

Figure 6-7 Setup for overnight biofilm assay.

II. Quantitatively analyze biofilm formation

Heavy biofilms retain more crystal violet than light biofilms (Fig. 6-6). Therefore, the amount of biofilm matter present can be quantified by measuring the amount of dye retained. The spectrophotometer you used to monitor cell density in Challenge Three will be used again in this challenge. Crystal violet is soluble in alcohol and absorbs light at a wavelength of 540 nm. You will solubilize the crystal violet in each well, transfer it to a cuvette, and record the absorbance.

PROCEDURE

Quantifying the amount of biofilm by spectrophotometry

1. Working with your partner, pipette 250 µl 95% ethanol into wells A2, A5, and A8, then transfer the contents of each well to a separate, labeled microcentrifuge tube. Repeat the wash with an additional 250 µl 95% ethanol.
2. Add 500 µl deionized water to each of the tubes and mix by inverting the capped tube several times.
3. Mix 500 µl deionized water with 500 µl 95% ethanol and place in a cuvette. This will be used as a blank for spectrophotometer.
4. Record the OD_{540} (optical density at 540 nm) of each sample in Table 6-1.

Table 6-1 Quantitative measurement of biofilms

Group	Isolate number	Absorbance (A_{540})		
		Isolate	*E. coli* (positive control)	*B. cereus* (negative control)

III. Determine whether the hospital strains produce quorum-sensing compounds

The effects of quorum sensing are often not easy to observe, and for that reason, genetically engineered reporter strains have been constructed. These strains respond to quorum-sensing compounds in a way that is easily detectable. The reporter in this experiment is a strain of the Gram-negative bacterium *Agrobacterium tumefaciens* (Fuqua and Winans, 1994). There are two plasmids in this strain. One of these contains the gene *traR* (Fig. 6-8A), a homolog of *luxR* (Fig. 6-2). The TraR protein can bind to a number of different AHL molecules and then activate the promoter of the *luxI* homolog *traI*. The other plasmid contains the *lacZ* gene under the control of the *traI* promoter. The *lacZ* gene encodes β-galactosidase, an enzyme that cleaves the glycosidic bond in lactose, generating galactose and glucose. X-Gal (5-bromo-4-chloro-3-indolyl-β-D-galactopyranoside) is a synthetic lactose analog consisting of galactose joined to a substituted indole ring by a glycosidic bond. X-Gal is colorless, but when the glycosidic bond is cleaved, a blue compound is produced (Fig. 6-8B).

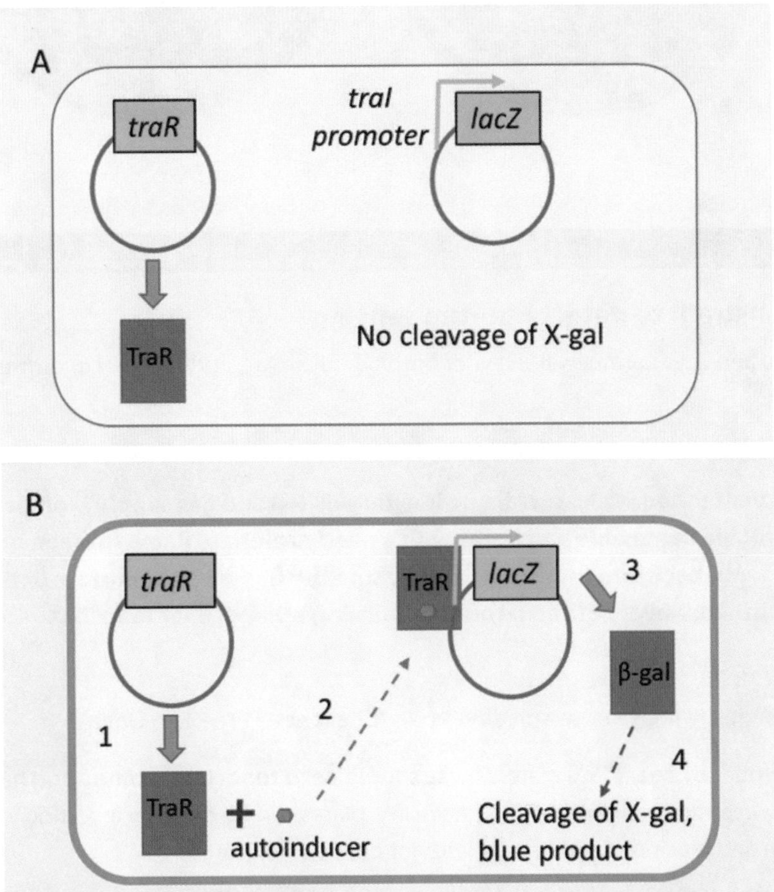

Figure 6-8 System for detecting quorum sensing with a reporter strain. The reporter strain contains a plasmid with the *traR* gene and a second plasmid with *lacZ* under control of the *traI* promoter. If the autoinducer produced by the test strain is active, after diffusing into the cell it will bind to the TraR protein. The TraR/autoinducer complex will then activate the *traI* promoter, inducing expression of *lacZ*. The β-galactosidase that is produced will then cleave the X-gal in the plate to produce a blue compound.

What happens when this reporter strain detects an AHL in the medium? The AHL binds to TraR, which in turn activates the *traI* promoter, and as a result, β-galactosidase is produced. If X-Gal is in the medium, it will turn blue in the presence of the enzyme. Thus, a quorum-sensing strain producing an active AHL is easily detected by the blue color of the reporter strain cells (Fig. 6-9).

Figure 6-9 Response of the *A. tumefaciens* reporter strain to a control strain producing AHL (left) and to a negative control strain (right).

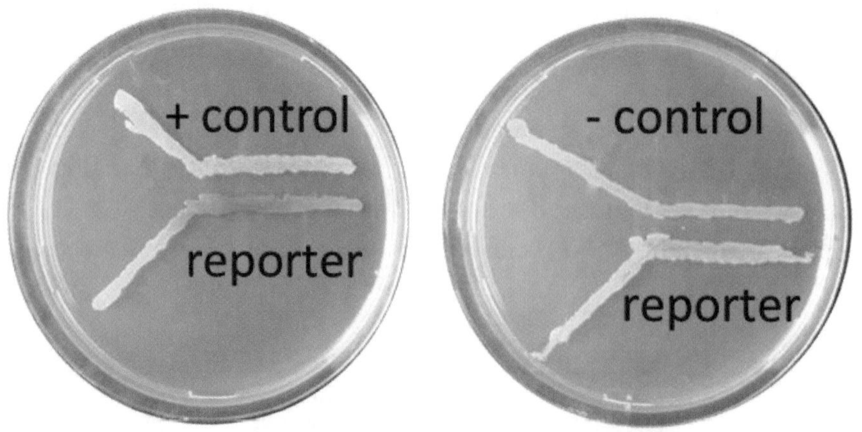

PROCEDURE

Using a reporter strain to detect quorum sensing

Working in pairs, you and your partner will test your hospital isolate for production of quorum-sensing molecules.

SAFETY

Never leave flame unattended. Make sure flame is extinguished and gas supply is off before leaving the lab. Ethanol is highly flammable and burns with a nearly colorless flame that can go unnoticed. Before placing the L-rod back into the ethanol, make sure the flame on the surface of the L-rod is completely extinguished. Never hold an L-rod with burning ethanol over the ethanol container.

Making the X-Gal plates

The technique is essentially the same as spreading cells (Challenge Three, Lab One).

1. Select two tryptic soy agar (TSA) plates. These should be at room temperature and have a dry agar surface (no puddles of moisture). If necessary, plates can be dried by inverting with the top partially open and placing in a 37°C incubator for about 15 minutes.
2. Deposit 100 µl of the X-Gal solution in the center of the plate. Using a sterile spreader, distribute the liquid across the surface of the plate. Continue spreading until *all* the liquid has been absorbed by the agar medium.

   Spread the solution shortly after depositing it on the plate: you don't want all the X-Gal to be absorbed at the site of the drop.

3. If you are using a glass spreader, follow the steps below:
   a. Dip the bent section of the spreader in alcohol, then touch the spreader to a gas flame to ignite the transferred alcohol. Wait until the alcohol has been consumed (just a few seconds). Do *not* hold the spreader in the flame.
   b. Let the spreader cool for 5 to 10 seconds. You can touch the flamed part to the surface of the agar (away from the X-Gal solution) to hasten cooling. Spread the X-Gal until all of the liquid has been absorbed.
   c. Reflame the spreader before placing on the bench.

Liquid cultures of the two hospital strains and the positive control strain will be available in the lab. Using a sterile loop, transfer cells from your hospital strain onto an X-Gal plate. Beginning in the upper left side of the plate, drag cells downward to near the middle of the plate, then straight across, as shown in Fig. 6-10.

Transfer cells from the liquid culture of the reporter strain to the plate. Start at the bottom left side of the plate (Fig. 6-10) and drag cells upward toward the middle, coming close to (but not touching!) the test strain, then straight across parallel to the test strain. The "Y" configuration of the streaks is to demonstrate the effect of diffusion: the farther away the two strains are, the less signal (i.e., AHL molecules) will reach the reporter. Repeat this procedure using the positive control strain and the reporter strain.

Incubate the plates at 30°C. In positive cases, it might take several days for the blue color to appear. Your instructor will store the plates for you to view during the next lab period.

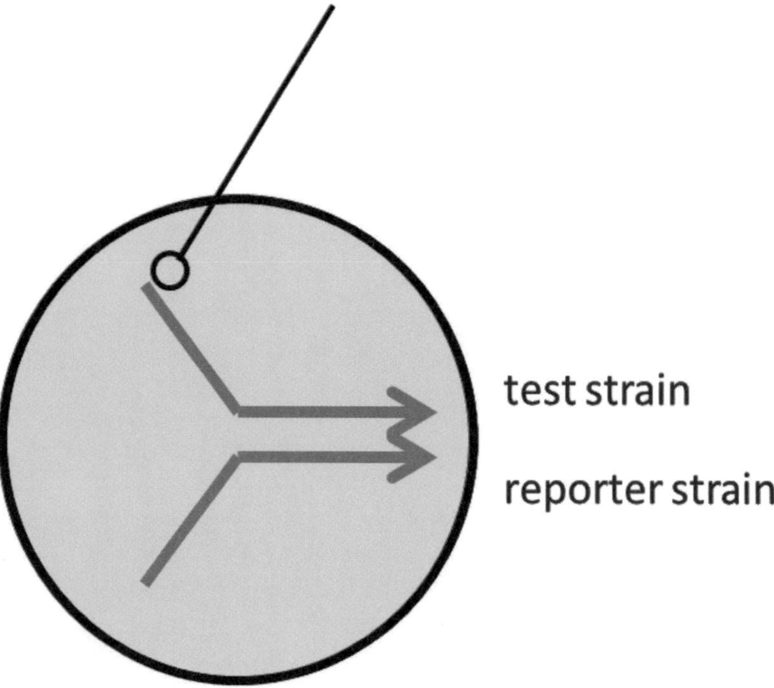

Figure 6-10 Streaking cells for the quorum-sensing assay.

QUESTIONS

Questions are designated B1 to B6 according to the six levels of Bloom's taxonomy.

1. Why are the wells rinsed with water before staining with crystal violet? (B2)
2. A strain of *Staphylococcus epidermidis* is tested with the *A. tumefaciens* quorum-sensing reporter strain. After overnight incubation, the positive control, but not the *S. epidermidis* strain, was blue. Can you make any conclusions based on this result? Why or why not? (B3)
3. The *A. tumefaciens* quorum-sensing reporter strain was also used to test a strain of *E. coli.* After incubation, the *E. coli* cells were blue but the reporter strain was not. What is the most likely explanation for this result? (B3)

Lab Two

BACKGROUND

Swimming

KEY POINTS

- Flagella are long filaments used by some bacteria, including *E. coli*, for motility.
- Swimming can result in random or directed movement. When the movement is directed toward a chemoattractant or away from a chemorepellant, it is called chemotaxis.
- Chemotaxis plays a role in pathogenesis: it is thought that strains of *E. coli* that cause urinary tract infections gain access to the bladder through chemotaxis to amino acids that are commonly found in urine.

Most strains of *E. coli* are motile in liquid, and are able to swim at a speed of about 3×10^{-5} meters/second, or roughly 10 cell lengths/second (Darnton et al., 2007; DiLuzio et al., 2005). Motility in *E. coli* is due to multiple flagella, which are distributed randomly around the cell surface. There are usually about five flagella per *E. coli* cell, although there can be as few as one and as many as fifteen (Cohen-Ben-Lulu et al., 2008). Each flagellum extends from a shaft, part of a molecular motor that rotates the shaft about 100 turns per second. A close inspection of a motile *E. coli* cell reveals that it switches between swimming and tumbling. When the bacteria are swimming, the flagella are being rotated in the counterclockwise direction. This direction of rotation causes the flagella to associate, forming a thick, corkscrew-like filament that propels the cell. Clockwise rotation results in dissociation of the flagella so that they are no longer functioning together, and the cell consequently tumbles in the medium without directed movement. Switching between swimming and tumbling can be extremely rapid and can occur about once every second or less.

Cycles of swimming and tumbling do not result in net movement in any direction. Bacteria swim for a short time in one direction, undergo tumbling, and then swim in a new direction determined by the orientation of the cell at the end of the tumble. Since this orientation is different and determined randomly, overall cells do not swim in any one direction but undertake a "random walk" through liquid.

When *E. coli* encounters a concentration gradient of attractant or repellant, then movement becomes directed: toward an attractant and away from a repellant. This behavior is called **chemotaxis** and is controlled by proteins encoded by the *che* genes. Chemotaxis happens because the cells sense the gradient and adjust their swim times in response. If a cell happens to be oriented toward an increasing concentration of attractant after tumbling, then the swim time is longer before the next episode of tumbling. In contrast, there is no increase in swim time if the cell is oriented so that it swims toward a lower concentration of attractant. The swim time is then reset to the shorter period and the same thing happens after the next round of tumbling. In this way, there is net movement toward the attractant (Fig. 6-11). A similar strategy is used if instead the cell is swimming in a gradient of repellant: cells swim for longer times in the direction of a decreasing gradient of repellant. Note that three conditions must be fulfilled for *E. coli* to engage in chemotactic behavior: the cells must be motile (that is, have functional flagella and enough energy to run the motor), there must be a *gradient* of an attracting or repelling substance, and the cells must be able to sense this gradient. *E. coli* senses a gradient temporally. While it is swimming, it "notices" whether the concentration of an attractant or repellant is increasing and responds accordingly.

Urinary tract infections result in an estimated 8 million doctor's office visits each year, and they are also the number one health care-associated infection reported in Europe and the United States (Schappert and Rechtsteiner, 2011; World Health Organization, 2011). Most of these infections are caused by uropathogenic *E. coli* (UPEC). UPEC can survive in the intestinal tract, as can other pathogenic strains of *E. coli*, including the important group of enterohemorrhagic *E. coli* (EHEC). EHEC strains produce a toxin that causes bloody diarrhea. The pathogen *E. coli* O157:H7, discussed in Challenge Four, is a member of the UPEC group. When

Figure 6-11 (Top) Cells do not show a chemotactic response to a single concentration of chemoattractant and are uniformly distributed in the medium. (Bottom) Cells swim for longer times in the direction of an increasing concentration of chemoattractant, resulting in net movement of the population toward the highest concentration.

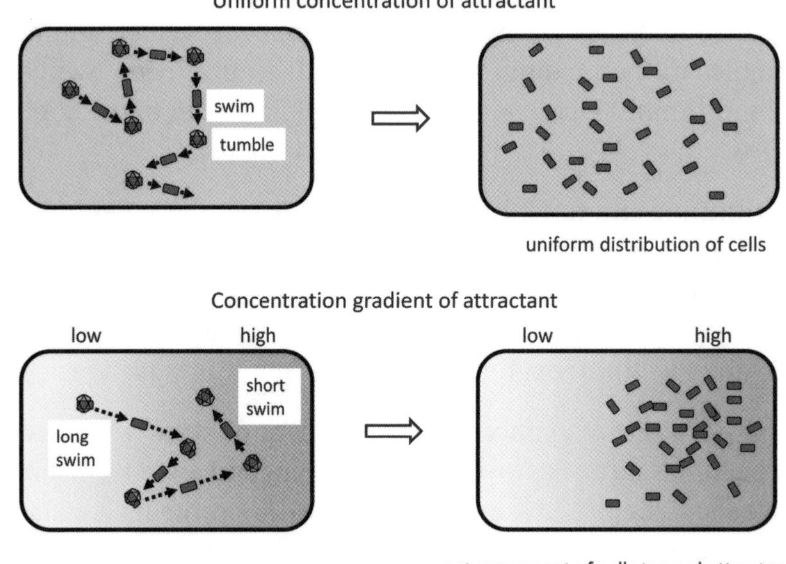

UPEC strains exit the gastrointestinal tract, they can then enter the urinary tract to cause an infection. In contrast, urinary tract infections caused by EHEC strains are exceedingly rare. UPEC strains are better suited for the urinary tract, in part due to their ability to utilize the toxic amino acid D-serine, abundant in the urine, as a carbon source (Bockmann et al., 1992). In EHEC strains, the genes required for catabolism of D-serine have been replaced with genes for sucrose utilization. The presence of the serine utilization genes confers a competitive advantage for UPEC strains in the urinary tract because while amino acids are present in the urine of healthy individuals, sugars are generally absent. Additionally, D-serine has been shown to repress the system used by EHEC to attach to host cells. If EHEC does enter the urinary tract, the high levels of D-serine downregulate this system and reduce the ability of EHEC to attach to host cells (Connolly et al., 2015).

Chemotaxis also plays a role in UPEC fitness. UPEC strains exhibit positive chemotaxis to urine, with amino acids acting as the strongest chemoattractants (Raterman and Welch, 2013). It is hypothesized that UPEC may use chemotaxis to gain access to the bladder (Raterman and Welch, 2013). Mutant UPEC strains deficient in motility and chemotaxis display decreased fitness relative to wild-type strains, supporting a role for chemotaxis in pathogenesis (Lane et al., 2005).

Lab Two

1. *Visualize biofilm formation using crystal violet.*
2. *Quantitatively analyze biofilm formation using spectrophotometry.*
3. ***Determine if the strains produce quorum-sensing compounds (continuing).***
4. ***Determine whether compounds in urine are chemoattractants for these strains.***
5. ***Analyze the effectiveness of chemical cleaners by the disk diffusion assay.***

Learning outcomes

After this lab, students will be able to:

- **a.** Interpret results of a quorum-sensing assay.
- **b.** Do a plug-in-soft-agar assay to test for chemotaxis.
- **c.** Perform and interpret results of a Kirby-Bauer disk diffusion assay.

I. Complete the analysis of quorum sensing

Observe the quorum-sensing plates from the previous lab session. Record your findings in Table 6-2 in the "Quorum sensing (Yes/No)" column. Does the reporter strain detect signaling from any of the hospital strains? The positive control will turn the reporter strain blue during this period. If one of the hospital strains also turns the reporter strain blue, the color might be less or more intense than for the control. Different strains produce different AHL molecules, and the reporter strain is not equally sensitive to all of these molecules.

Table 6-2 Examining hospital isolates for quorum sensing and chemotaxis

Student pair	Isolate number	Quorum sensing: (Yes/No)	Chemotaxis Attractant (A) Repellant (R) No response (N)				
			H_2O	leucine	aspartate	serine	acetate

II. Assay the hospital strains for chemotaxis to different compounds

In Challenge One, you used soft agar plates to determine which strains were motile. Cells swim outward from the site of inoculation because as the nutrients in the medium are used they become locally depleted and a concentration gradient is formed. However, this test does not tell us which chemicals in the medium are acting as attractants. Most health care-associated urinary tract infections are the result of contaminated catheters, and your hospital isolates have been found on catheter trays. Therefore, you want to determine if these strains respond chemotactically to common amino acids found in urine. You will use the "plug-in-soft-agar" assay to test four different compounds. In this assay, minimal medium-soft agar is used. The medium contains salts and 0.2% glucose as a carbon source. As before, cells will be stabbed into the center of the plate. In addition, a plug of hard agar containing the test chemical will be inserted into the plate, several millimeters from the edge, as shown in Fig. 6-12. The chemical will diffuse from the plug into the soft agar, thus establishing a gradient. If the chemical is a chemoattractant, growth will be seen only in the direction of the gradient (Fig. 6-12).

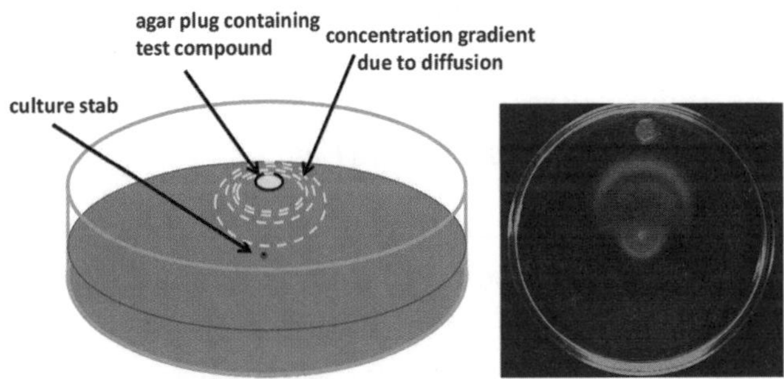

Figure 6-12 (Left) A hard agar plug containing a chemical is inserted into a soft agar plate. A culture is stabbed into the middle of the plate and incubated overnight. (Right) Cells exhibiting positive chemotaxis to a chemoattractant.

PROCEDURE

Testing for chemotaxis with the "plug-in-soft-agar" assay

Working in pairs, obtain your assigned hospital isolate for the assay. You will test the four compounds serine, leucine, sodium aspartate, and sodium acetate, as well as a negative control (water).

1. Obtain five minimal medium-soft agar plates. Attach a sterile cut tip (provided by your instructor) to a micropipette. Make a hole in the soft agar medium, several millimeters from the edge of the plate, by pressing down into the medium with the tip and then gently drawing up the soft agar with the micropipette. Discard the soft agar from the tip. Do this for each of the five plates.
2. Sets of five hard agar (1.4%) plates will be provided in the lab for shared use. Four of these will contain one of the test compounds at a concentration of 10 mM in the agar. The fifth plate will contain agar without any addition. Using a sterile cut tip as before, draw up a plug of the hard agar

Procedure continues on next page

Testing for chemotaxis with the "plug-in-soft-agar" assay (*continued*)

from each plate and insert the plug into the hole in one of the soft agar plates. The plug can be expelled by pressure on the tip plunger, as though expelling liquid. If it resists coming out, then remove the tip and use a second, unmodified sterile tip to gently push the plug into the hole. Prepare a plate for each of the test compounds. Make sure each plate is labeled on the cover with the test compound or designated as the control. You will not be able to invert the plate for labeling.

3. After the plates are ready, stab your hospital strain from a liquid culture (provided) into the center of the plate. Incubate the plates at 30°C with the cover side up.
4. Observe the plates and record the results after 24 to 48 hours. Incubate the plates further if only faint growth is observed. Categorize the compounds according to whether they are an attractant, repellant, or neither and record your conclusion in Table 6-2.

III. Determine the effectiveness of chemical cleaners

Disk diffusion assays are frequently used to determine the sensitivity of a strain to different antibiotics (Challenge Three). They can be used equally well to assess the antimicrobial activity of different disinfecting agents. A sterile paper disk is saturated with a chemical of interest and placed onto a plate swabbed with culture. The chemical will diffuse outward into the medium, resulting in a gradient. A clear zone of inhibition will appear where the chemical reaches bactericidal concentrations. Effective chemicals will result in a large zone of inhibition (Fig. 6-13).

You and your partner will determine the susceptibility of your assigned hospital isolate to various disinfecting agents. You may choose up to four different chemicals, or you may choose to make four dilutions of one chemical (for example, 0.1, 1, 5, and 10% bleach solutions). Your instructor will provide you with a list of available chemicals. If there is a specific disinfectant agent you would like to try, you will need to request approval from your instructor in advance.

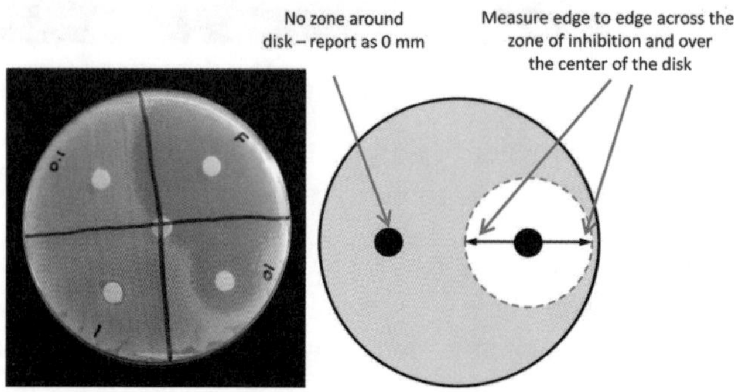

Figure 6-13 The Kirby-Bauer assay. (left) Starting with the top right quadrant and moving clockwise, disks were soaked with 100% bleach, 10% bleach, 1% bleach, and 0.1% bleach. The water control disk is in the center. A zone of inhibition produced by 100% bleach extends into the neighboring quadrants. (right) Place a ruler over the center of the paper disk to measure the diameter of the zone of inhibition. Figure adapted from ASM Protocol, 2009. Protocol 3189. Credit: Jan Hudzicki.

PROCEDURE

Testing for chemical effectiveness with the Kirby-Bauer disk diffusion assay

1. Label a TSA plate with your name, the date, and the name of your assigned hospital isolate. Divide the plate into quadrants and label each quadrant with the name and concentration of the cleaning solution you are testing.
2. An overnight culture of your hospital isolate will be available in the lab. Using a sterile cotton swab, sample the culture and then spread on the plate, making sure to cover the entire surface evenly. To ensure proper coverage, it is best to swab the plate from top to bottom, then turn the plate 90° and swab from top to bottom again. It may be necessary to resample the culture before you turn the plate.
3. Use an incinerator to sterilize forceps. If an incinerator is not available, dip the grasping end of the forceps into alcohol and ignite by passing through a flame to sterilize. Do not hold the forceps in the flame.
4. Using the forceps, pick up a sterile paper disk, dip the disk into sterile water, and place in the center of the plate. This will be the control. Re-sterilize the forceps, then wet a sterile paper disk with one of your cleaning solutions and apply to the center of a labeled quadrant. Repeat for the other three solutions you are testing.
5. Incubate the plate overnight at 30°C with the *cover up, agar side down* to prevent the paper disks from falling off.

Return the following day to observe. Use a ruler to measure the diameter of the zone of inhibition as shown in Fig. 6-13 (right). Record your results in Table 6-3 and rank the solutions on each disk according to their effectiveness in preventing bacterial growth.

QUESTIONS

1. Why is minimal medium rather than TSB used to make up the soft agar plates for the chemotaxis assay? (B2)
2. A swim plate was set up, but a plug containing repellant was placed in the center and cells stabbed at a point midway between the edge of the plate and the plug. Sketch a picture depicting the appearance of the population of swimming cells in that case. (B3)
3. A simple point mutation (C to T) in one of the genes for chemotaxis in *E. coli* resulted in cells that could no longer recognize a gradient of attractant and, as a result, could no longer swim. A microbiologist irradiated the cells with UV in order to get revertants (cells with the reverse mutation, T to C). How could such revertants be easily isolated? (B6)
4. Design an experiment to show that the motile, photosynthetic organism *Rhodospirillum rubrum* is positively phototactic (migrates in the direction of increasing light). Include positive and negative controls, as well as variables you would measure. (B6)

Table 6-3 Sensitivity of hospital isolates to disinfectants

Group and isolate number	Disinfectant or control	Conc. (100% = full strength)	Zone of inhibition (mm)	Effectiveness rank 1 (best) → 5 (worst)

Solving Challenge Six

Using the information from Tables 6-1, 6-2, and 6-3, compare your results to those of the other groups. As a class, complete Table 6-4. What are the differences between the two hospital strains? Based on the class results, can you recommend a disinfectant that will be effective against both strains?

Table 6-4 Summary of the properties of the two hospital isolates and recommendations

	Biofilm producer(Y/N)	Ability to Quorum Sense(Y/N)	Positive chemotaxis(chemicals)	Recommended disinfectants
Hospital isolate 1				
Hospital isolate 2				

BIBLIOGRAPHY

Bockmann J, Heuel H, Lengeler JW. 1992. Characterization of a chromosomally encoded, non-PTS metabolic pathway for sucrose utilization in *Escherichia coli* EC3132. *Mol Gen Genet* 235:22–32. http://www.ncbi.nlm.nih.gov/pubmed/1435727.

Case RJ, Labbate M, Kjelleberg S. 2008. AHL-driven quorum-sensing circuits: their frequency and function among the Proteobacteria. *ISME J* **2:**345–349. http://www.nature.com/doifinder/10.1038/ismej.2008.13.

Cohen-Ben-Lulu GN, Francis NR, Shimoni E, Noy D, Davidov Y, Prasad K, Sagi Y, Cecchini G, Johnstone RM, Eisenbach M. 2008. The bacterial flagellar switch complex is getting more complex. *EMBO J* **27:**1134–1144. http://www.pubmedcentral.nih.gov/articlerender.fcgi?artid=2323253&tool=pmcentrez&rendertype=abstract.

Connolly JP, Goldstone RJ, Burgess K, Cogdell RJ, Beatson SA, Vollmer W, Smith DG, Roe AJ. 2015. The host metabolite D-serine contributes to bacterial niche specificity through gene selection. *ISME J* **9:**1039–1051. https://www.ncbi.nlm.nih.gov/pmc/articles/PMC4366372/.

Darnton NC, Turner L, Rojevsky S, Berg HC. 2007. On torque and tumbling in swimming *Escherichia coli*. *J Bacteriol* **189:**1756–1764. http://jb.asm.org/content/189/5/1756.long.

Darwish SF, Asfour H A. 2013. Investigation of biofilm forming ability in *Staphylococci* causing bovine mastitis using phenotypic and genotypic assays. *ScientificWorldJournal* **2013:**378492. http://www.ncbi.nlm.nih.gov/pubmed/24298212.

DiLuzio WR, Turner L, Mayer M, Garstecki P, Weibel DB, Berg HC, Whitesides GM. 2005. *Escherichia coli* swim on the right-hand side. *Nature* **435:**1271–1274. http://www.nature.com/doifinder/10.1038/nature03660.

Fu W, Forster T, Mayer O, Curtin JJ, Lehman SM, Donlan RM. 2010. Bacteriophage cocktail for the prevention of biofilm formation by *Pseudomonas aeruginosa* on catheters in an in vitro model system. *Antimicrob Agents Chemother* **54:**397–404. http://www.ncbi.nlm.nih.gov/pubmed/19822702.

Fuqua WC, Winans SC. 1994. A LuxR-LuxI type regulatory system activates *Agrobacterium* Ti plasmid conjugal transfer in the presence of a plant tumor metabolite. *J Bacteriol* **176:**2796–2806. http://jb.asm.org/content/176/10/2796.long.

Lane MC, Lockatell V, Monterosso G, Lamphier D, Weinert J, Hebel JR, Johnson DE, Mobley HL. 2005. Role of motility in the colonization of uropathogenic *Escherichia coli* in the urinary tract. *Infect Immun* **73:**7644–7656. http://iai.asm.org/content/73/11/7644.long.

McFall-Ngai M, Hadfield MG, Bosch TC, Carey HV, Domazet-Lošo T, Douglas AE, Dubilier N, Eberl G, Fukami T, Gilbert SF, Hentschel U, King N, Kjelleberg S, Knoll AH, Kremer N, Mazmanian SK, Metcalf JL, Nealson K, Pierce NE, Rawls JF, Reid A, Ruby EG, Rumpho M, Sanders JG, Tautz D, Wernegreen JJ. 2013. Animals in a bacterial world, a new imperative for the life sciences. *Proc Natl Acad Sci U S A* **110:**3229–3236.

Metwalli KH, Khan SA, Krom BP, Jabra-Rizk MA. 2013. *Streptococcus mutans*, *Candida albicans*, and the human mouth: a sticky situation. *PLoS Pathog* **9:**e1003616. http://dx.plos.org/10.1371/journal.ppat.1003616.

Monroe D. 2007. Looking for chinks in the armor of bacterial biofilms. *PLoS Biol* **5:**e307. http://dx.plos.org/10.1371/journal.pbio.0050307.

Raterman EL, Welch RA. 2013. Chemoreceptors of *Escherichia coli* CFT073 play redundant roles in chemotaxis toward urine. *PLoS One* **8:**e54133. http://dx.plos.org/10.1371/journal.pone.0054133.

Schappert SM, Rechtsteiner EA. 2011. Ambulatory medical care utilization estimates for 2007. *Vital Health Stat* **13:1–38.** http://www.ncbi.nlm.nih.gov/pubmed/21614897.

Waters C, Bassler B. 2005. Quorum sensing: cell-to-cell communication in bacteria. *Annu Rev Cell Dev Biol* **21:**319–346. http://www.annualreviews.org/doi/full/10.1146/annurev.cellbio.21.012704.131001.

World Health Organization. 2011. *Report on the Burden of Endemic Health Care-Associated Infection Worldwide.* World Health Organization, Geneva, Switzerland. http://apps.who.int/iris/bitstream/10665/80135/1/9789241501507_eng.pdf.